The Engineering Design Process

Barry Hawkes AMIED
Lecturer in Engineering Design and Mathematics
Canterbury College of Technology

Ray Abinett TEng, MIED
Principal Design Engineer, Lucas CAV Ltd.
Lecturer in Engineering Design, Mid-Kent College
of Higher and Further Education, Chatham

 LONGMAN

Addison Wesley Longman Limited
Edinburgh Gate, Harlow,
Essex CM20 2JE, England
and Associated Companies throughout the world

First published 1984
Reprinted 1985
Reprinted by Longman Scientific & Technical 1986, 1989, 1990, 1992, 1993
Reprinted by Addison Wesley Longman Limited 1997

ISBN 0 582 99471 3

Produced by Longman Singapore Publishers Pte Ltd
Printed in Singapore

Contents

Acknowledgments

The authors wish to express particular thanks to The Ford Motor Company and to Benford Ltd. (Warwick) for substantial contributions to sections of this book.

The authors wish to thank the following companies, organisations and individuals for reference to their products and design research:

The Architects Journal
B.L. Ltd.
B.O.C. Ltd.
Chas. A. Blatchford Ltd.
Bridgeport Textron Ltd.
Brockhouse Harvey Frost Ltd.
Canterbury College of Technology
The Coalbrookdale Co. Ltd.
Compenda Ltd.
Computer-aided Design Centre (Cambridge)
Computational Mechanics Centre
Computervision Corporation
Delta Computer-aided Engineering
E.C.S. Ltd.
Engineering Computer Services Ltd.
Engineering Designer
Fisher Price Toys Ltd.
Flymo Ltd.
Frister & Rossman Ltd.
Mr. M. J. Gornal (inventor)
G. S. Iona Ltd.
Hitachi

The Institution of Engineering Designers
Karpark Ltd.
Kenwood (Thorn EMI) Ltd.
Lesney Industries Ltd.
William Levene Ltd.
The Machine Tool Industry Research Association (MTIRA)
Medway College of Design
Mondiale Ltd.
The National Engineering Laboratory (NEL)
Oxfam
PAFEC Ltd.
Philips Electronics Ltd.
Preston Polytechnic
Renault UK Ltd.
Ruston & Hornsby Ltd.
Russell Hobbs Ltd.
Shandon Southern Products Ltd.
Mr. C. B. Suresh Babu (design engineer)
Tektronix
Vickers Ltd.

Preface

The central theme of this book is the presentation of engineering design as a fully-integrated process which controls each progressive stage in the creation of a new product. The ideal process must inevitably involve a systematic approach which allows maximum scope for innovative flare and takes full account of new technologies and changing social trends. Tackled this way, engineering design becomes a fresh and exciting subject which is completely relevant to the needs of modern society.

The term "engineering design" could, of course, cover a vast range of specialisms, from the design of complete process or power plants to integrated circuit design. In order to emphasise basic principles, we have concentrated mainly on mechanical design applied to product design in the context of consumer goods and small-to-medium capital outlay.

Building on the basic groundwork provided in *Engineering Design for Technicians* (by the same authors and publishers), this new book, therefore gives a total view of the process for aspiring product designers undertaking courses at higher technician or first-year degree level. It provides full coverage of the B/TEC units Level IV and V "Engineering Design" but includes many additional topics relevant to college, polytechnic, and university design courses. It is hoped that this book will also provide a useful reference for students taking a GCE A-level in design technology and for those studying engineering management.

The essential ingredients of the systematic approach are shown to be, first, the establishment of the primary need for a new product and to continue with particular emphasis on the need for well-detailed design specifications, on logical thought processes such as brain-storming and lateral thinking, and on the importance of evaluating a variety of possible solutions to a problem. The systematic procedures can only be successful if accompanied by efficient design organisation structures and planning techniques which are discussed and applied in some detail. Cost considerations and market competition dictate the need to produce designs which are both functionally sound and economically viable. Economic aspects have therefore been covered, including methods of reducing costs by modifying design form and standardising components.

The analytical content lays emphasis on: systematic procedures of dimensioning and tolerancing; the application of mathematical formulae to practical design problems; comparison of strength and fatigue characteristics for various design forms; and the importance of scale models to the design process. Several industrial and college case studies are included.

The increasing importance of the human aspect of engineering design is recognised with a chapter devoted to basic ergonomic considerations and ways of using anthropometric data to produce designs which give maximum comfort and efficiency to the user. Current trends such as the wider range of domestic appliances indicate a need for functional products which are also pleasing to the eye. Aspects of aesthetics are therefore discussed with several case studies showing how to make the appearance of a design appropriate to its function, the user, and the prevailing styles and fashions. Other case studies take due regard of growing social influences such as developments in medical equipment, D.I.Y. products, and intermediate technology in the developing world.

The essential role of new technologies in the design process is covered with detailed descriptions of techniques and applications in computer-aided drawing and design.

To underline the importance of inventive talent, a series of modern design innovation case studies is included.

Finally, the whole process is brought together with the description of a complete design case study of an actual product, followed by a series of complete design assignments.

B. R. Hawkes
R. E. Abinett

Maidstone/Rochester 1984

B.R.H. To Gill for her encouragement and endless patience throughout the whole project, Albert and Dianne for their help, and Georgia and Robert for their stolen time.

R.E.A. To Barbara for many hours of tireless effort in typing the manuscript, Brian Lawrence for his practical comments on design planning, and Suzanne, Tracey, and Jane for their patience.

1 Systematic Approach to Design

1.1 Outline of Approach

The purpose of this chapter is to emphasise the essential aspects of a systematic approach, to define common associated terminology, and to give some simple applications. In effect, the chapter provides the basic building block for further reading.

The principal role of the designer is that of *originator* in the process of creating new products. The initial task in this role is to identify the *primary need* for the product. The primary need is really the design problem expressed in clear terms. In addition, *secondary needs* must be established. These will be discussed later.

The Primary Need

The primary need can arise from various sources. It could be that the user of an existing piece of machinery requires to update the process and therefore modify the mechanism. Another reason could be exploitation of new technology. Further, a product may be in need of advancement over a competitor. Legislation changes could also result in the need to update or re-design a piece of equipment.

It is important to remember that the requirement as stated may not be the *true* need. For example, a market survey may show that a new machine is required to test vehicles under fully loaded conditions. Further investigations, however, may reveal that tests at only part-load are required. Considerable cost reductions in test equipment could therefore be achieved.

The Solution Process

There is no easy way of solving a design problem or of originating a design solution. However, there are certain rules which may be applied in order to reduce the effort. By a systematic process of elimination, workable design solutions may be evolved. This method is called *Systematic Design Technique*. Fig. 1.1 shows the process that should be adopted in order to achieve a design solution.

Analysis

When confronted with a design problem, the designer requires a clear train of thought and will therefore develop and establish an analysis. This is a form of stocktaking and will provide the answer to
  a) Method of operation
  b) Production procedures
  c) Method of power flow
  d) Method of control.
Analysis requires the application of both practical and theoretical knowledge. This may include
  a) Consideration of manufacturing tolerances
  b) Design form (shape of component)

Fig 1.1 The design
solution process

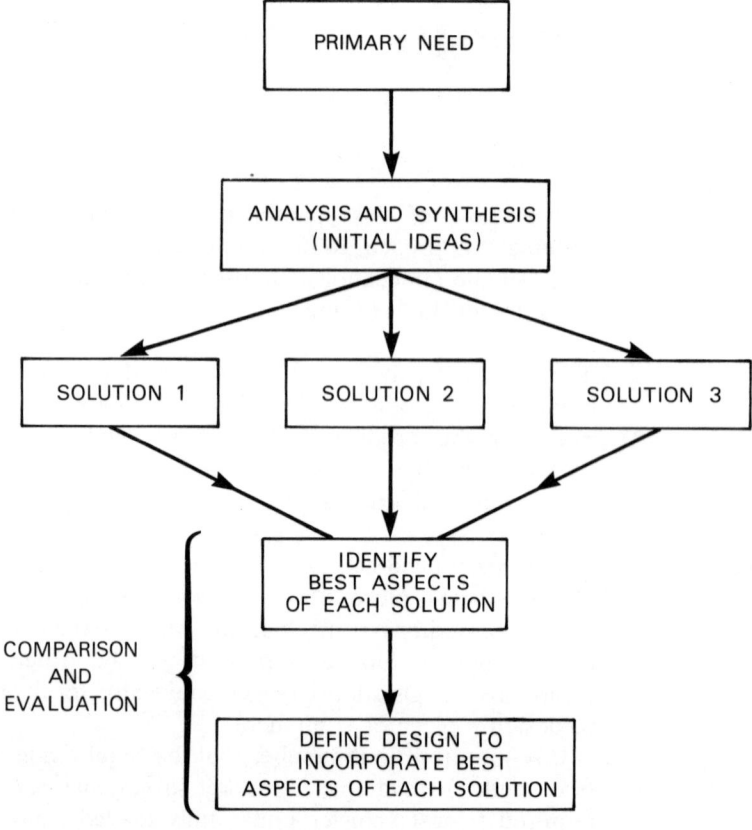

c) Ease of maintenance
d) Ergonomics
e) Strength considerations (e.g. stress analysis)
f) Size requirements.

**Synthesis,
Comparison and
Evaluation**

Synthesis involves collecting together design information and ideas. This
results in a number of alternative proposals being sketched and analysed. It
is recommended that a minimum of three schemes are produced so that
detailed comparisons may be made. The design sketches should be carefully
annotated to indicate important features. A systematic process of comparison and evaluation may now be undertaken by using a points system or
ranking technique.

Because of the many features that a design must conform to, and of the
need to operate under varying environmental conditions, a method of rating
is used to check out the design features of each idea. These features could
include

a) Fulfilment of function   f) Ease of maintenance
b) Reliability   g) Ease of manufacture
c) Cost   h) Efficiency of operation
d) Serviceability   i) Simplicity of layout.
e) Life

Ideally the method of evaluation should be kept relatively simple, so that time is not actually wasted on interpreting the results of the rating method. Conversely, the method should not be open to abuse whereby the results are manipulated to give the advantage to a favoured design—a mistake which is all too easily made.

**1.2
Example
of Approach**

By way of an example consider the case of a machine which operates a mechanism to move a table a certain distance and return it in one revolution of the drive motor. The purpose is to reject bottles from a conveyor.

Fig. 1.2 shows a diagramatic representation of the problem. It is required to design a simple mechanism to achieve this motion. As stated above, design sketches should be made and suitably annotated to show the important features of each design (Figs. 1.3–1.6).

Fig 1.2

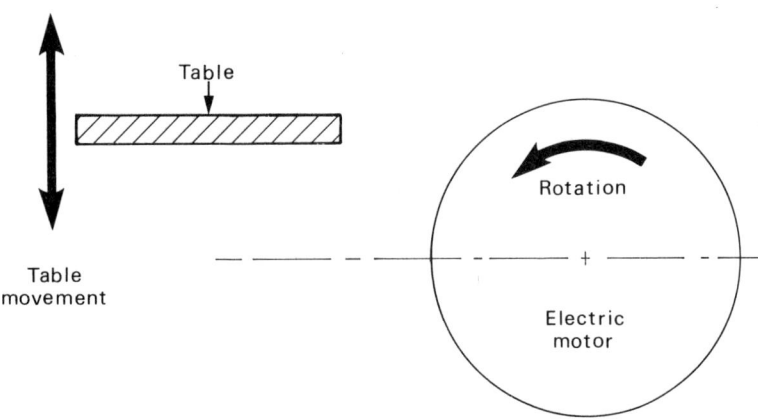

Fig 1.3

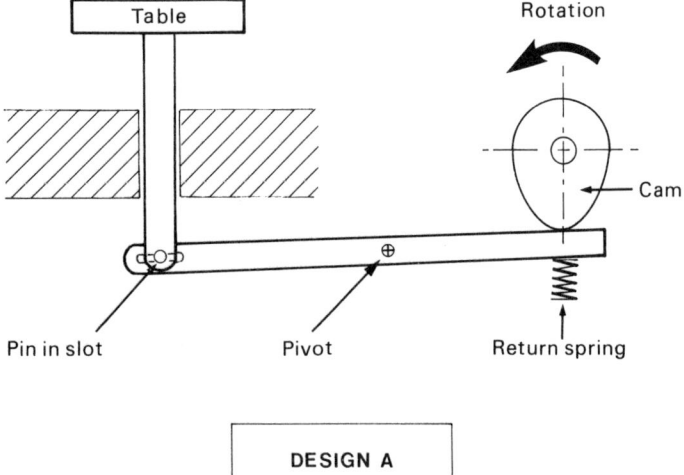

DESIGN A

Fig 1.4

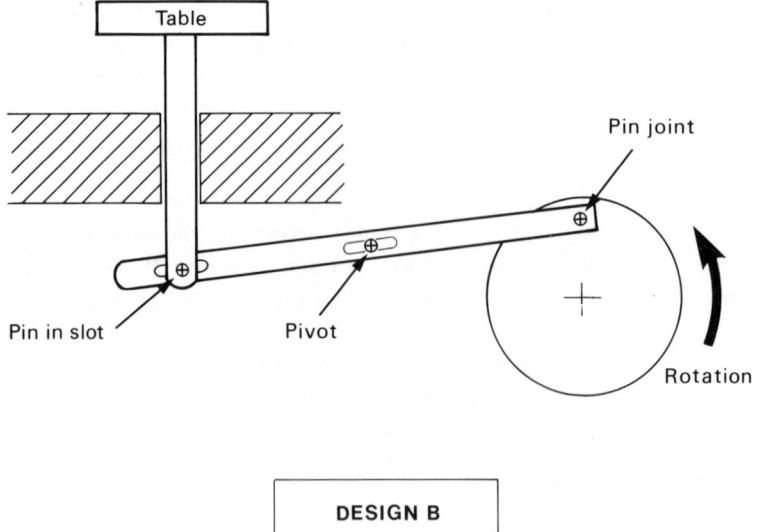

DESIGN B

Fig 1.5

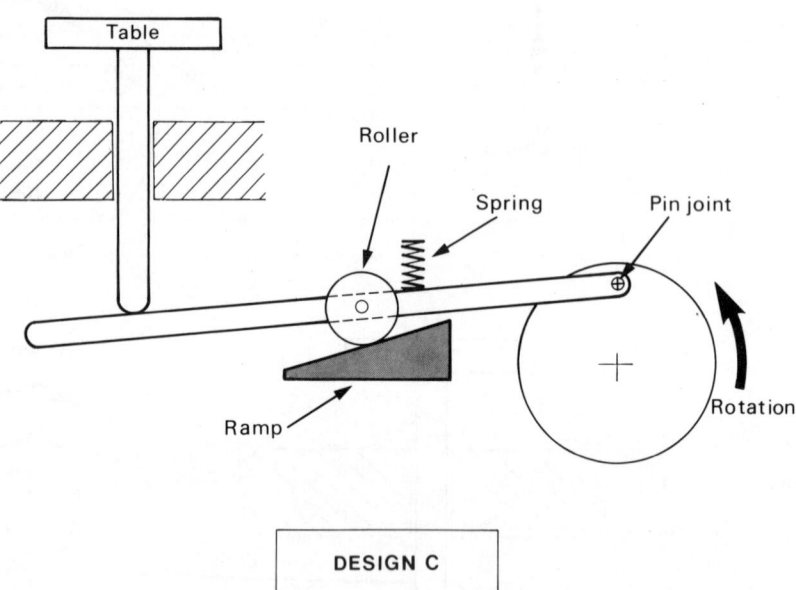

DESIGN C

Fig 1.6

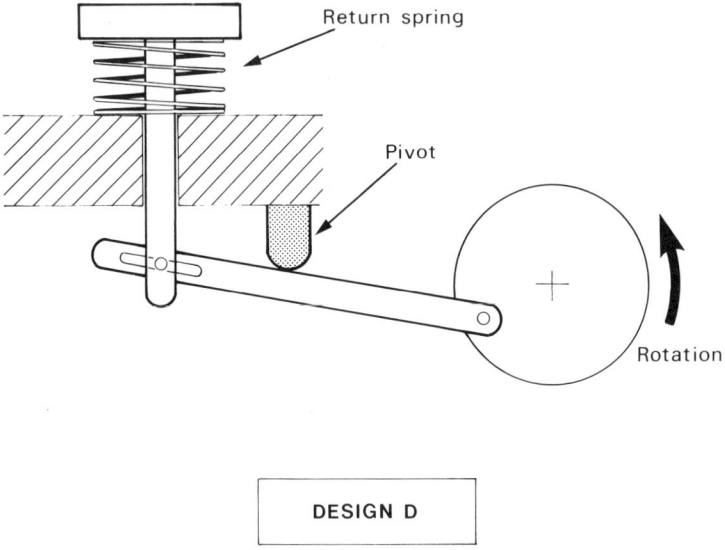

DESIGN D

Table 1.1 Design feature ratings

Feature	DESIGN A	DESIGN B	DESIGN C	DESIGN D
Fulfilment of function	High	High	High	High
Reliability	Medium	High	Low	Low
Serviceability	Medium	Medium	Medium	Low
Life	Low	Medium	Low	Low
Ease of maintenance	Medium	Medium	Medium	Low
Ease of manufacture	Medium	High	Low	Low
Efficiency of operation	Low	Medium	Low	Low
Simplicity of layout	Medium	High	Low	Low
Cost	Moderate	Cheap	Moderate	Expensive

High indicates the most satisfactory solution.
Medium indicates a nearly satisfactory solution.
Low indicates the unsatisfactory solution.

Table 1.2 Rating scores for each design

	DESIGN A	DESIGN B	DESIGN C	DESIGN D
High	1	4	1	1
Medium	5	4	2	0
Low	2	0	5	7
Cost	Moderate	Cheap	Moderate	Expensive

Rather than using a points system to identify the advantages of each of the items a) to i), Table 1.1 depicts a method of comparison based on High, Medium and Low ratings of each design feature considered.

This method clearly depends on the designer's view of the terms High, Medium and Low rating and they should be used objectively. It mostly prevents the designer from nominating a preferred design from the various sketches alone.

It can be seen that design B has more ratings in the high/medium category than the other designs and would therefore be considered the most suitable. Cost is favourable due to the simple design of the components used. The proposed design D has the lowest number of high/medium ratings and high cost (expensive). Therefore this design is considered to give most problems in manufacture, operation and cost.

When allocating a rating to each item, whether by this method or any other, the judgement should be based on the designer's personal experience of the space requirements for each part and the critical effect of the overall size upon its finished shape (form).

The method used above is only an example of what may be adopted to simplify choosing the best design based on the feature criteria required. Larger and more complex designs may require a more detailed analysis, broken down into individual assemblies and analysed in a similar manner.

1.3 Iterative Design Procedure

An iterative mathematical procedure is one in which an approximate solution to a problem is initially guessed and then fed into an iterative formula which reveals a more accurate solution. The improved solution is then put through the same procedure to reveal an even better solution, and the process is continued until a solution of the required accuracy is achieved. The overriding principle is that the error decreases with every successive solution.

A systematic design/re-design procedure must inevitably form a similar pattern to such mathematical processes. The iterative design procedure makes the realistic assumption that even the best design concepts may have to be modified for improvement at various stages in their development. With complex components, the modified versions may need further improvement until the ideal solution is achieved. An efficient iterative process will ensure

Fig 1.7 Iterative
design procedure

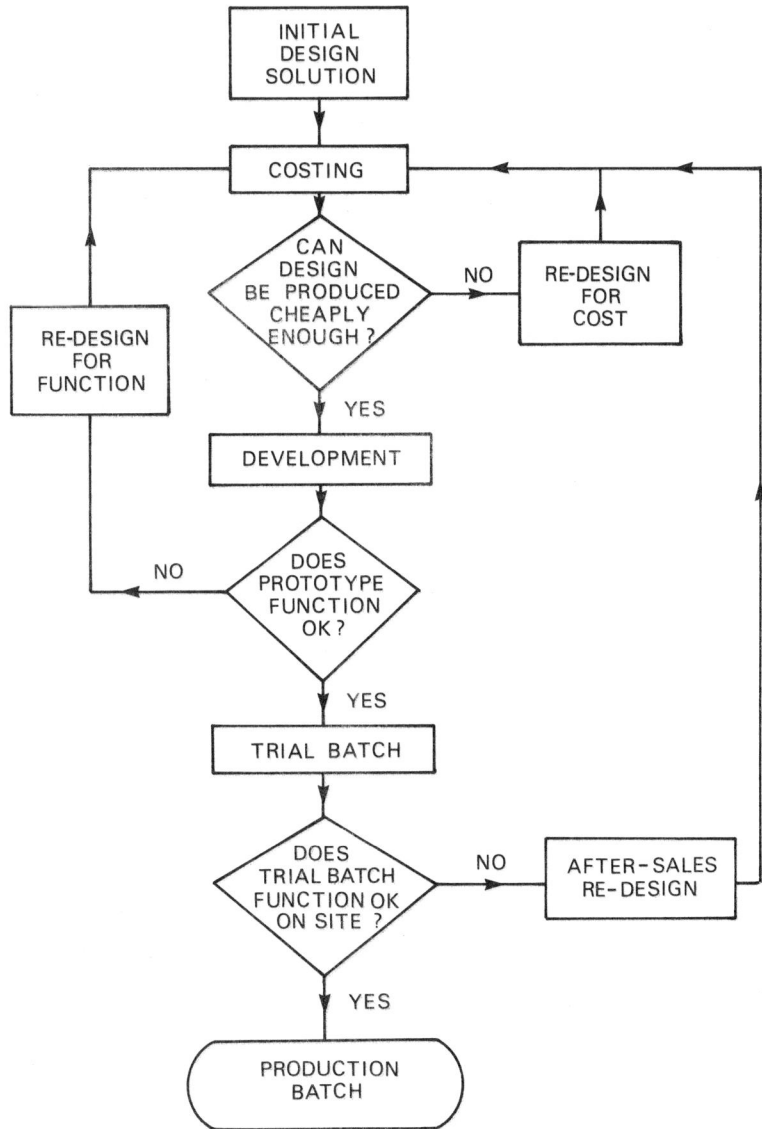

that each successive modification is less involved than the previous one. Fig.
1.7 illustrates the iterative design procedure in flow diagram form with
stages of re-design shown as feedback loops in the system.

Of course, it is essential that any stage of re-design be limited, where
possible to minor modifications, and, to ensure this, they must be tackled at
the earliest opportunity. Good liaison between departments is a critical
factor here. For example "re-design for cost" could include no more than
switching to a different supplier for a bought-out item after liaison with the
buying department. "Re-design for function" may involve no more than a

simple alteration of a design sketch after tentative development tests have been conducted. "After-sales re-design" can be tackled most effectively by obtaining feedback of site performance from a trial batch of the new design, supplied to a limited number of established customers. An example of this is discussed in Chapter 10.

1.4 Specification

The instruction to the designer to produce a solution comes through a specification. This may be supplied either by the customer or by a marketing request.

The designer could invent a specification but this is highly dangerous as the designer may not be aware of the market trends prevailing at the time. If an acceptable specification cannot be supplied, then it is better to make a judgement only after discussion with sales or marketing department or a consultant. A copy of the prepared draft specification should be given to them for comments or approval before commencing a design.

The British Standards recommendations for the preparation of a specification are contained in their published document PD 6112. This contains a list of items which should be included. However, some items may not be relevant to every design and the spec. writer must therefore select only those headings which are important.

Using the example of the bottle reject table, the specification for it may be written as follows using BS PD 6112 as a reference.

1.0.0 *Title* Mechanism for moving table suitable for rejecting glass bottles from a conveyor system.

2.0.0 *History & Background Information* It has been found necessary to introduce a mechanism on the special-purpose machine which loads bottles onto a conveyor. Because of imperfect sections in the glass itself, it was found that the reject bottles could be identified by using a light-refraction method (i.e. shining a light through the glass).

In order to reject the bottle a platform is required to push the bottles from the conveyor.

Preliminary discussion with the works engineering department has established that this is feasible provided that the mechanism works independently from the main conveyor drive.

3.0.0 *Scope of Specification*
3.1.0 The design must incorporate a time delay device which activates the mechanism drive motor. The purpose of this is to ensure that the reject bottle is aligned correctly with the moving table.
3.2.0 The table must be manufactured from materials able to prevent premature breakage of the bottle on the conveyor.

4.0.0 *Definitions*
4.1.0 Table: flat surface for pushing reject bottle from conveyor.
4.2.0 Cam Disc: eccentric or concentric pinned plate attached to the motor drive shaft.

5.0.0 *Conditions of Use*
5.1.0 The mechanism must be capable of moving forward a distance of 75 mm and returning to the same position after rejecting a bottle.
5.2.0 The movement of the table must not infringe the path of adjacent bottles passing along the conveyor.
5.3.0 There must be no environmental health hazard due to any feature of the mechanism or bottle rejection.

6.0.0 *Characteristics*
6.1.0 The shaft supporting the table must be mounted in a serviceable plain bearing having a length-to-diameter ratio of at least 2.5 : 1.
6.2.0 The mass of the moving parts must not impair the operation of the table assembly, and must not move forward at a speed greater than 0.125 m/sec.
6.3.0 All pivot points are to have oilite bearings.

7.0.0 *Reliability* The mechanism should be capable of operating for at least 4 years at a cycle rate of 1000 rejects per day.

8.0.0 *Servicing Features*
8.1.0 The mechanism must be capable of being serviced without having to touch the conveyor drive system.
8.2.0 Minimum lubrication of bearings and shafts is essential.
8.3.0 Easy access to the mechanism must be possible when safety guards are fitted.

1.5 Secondary Needs

The primary needs in relation to design activity have been discussed in section 1.1. Once these are established, there are other needs to be satisfied. These come under the heading of secondary needs and may consist of the following:
    a) Reliability (further)
    b) Ergonomics and Anthropometrics
    c) Aesthetics
    d) Safety
    e) Economics.
These subjects will be covered in later chapters.

1.6 Design Life Cycle

A design generally goes through a life cycle from the point of initiation, until the need arises to re-design it in line with the latest market requirements. Fig. 1.8 shows a typical graph of how this could occur. Just like people, certain "ages" of design can be identified on this graph.

1.7 Invention and Lateral Thinking

Pure inventive geniuses are few and far between. Normally an idea is sparked by an occurrence, which then initiates a design concept. This idea may have lain dormant in the brain for a long time. There are methods of

Fig 1.8 Typical design
life cycle

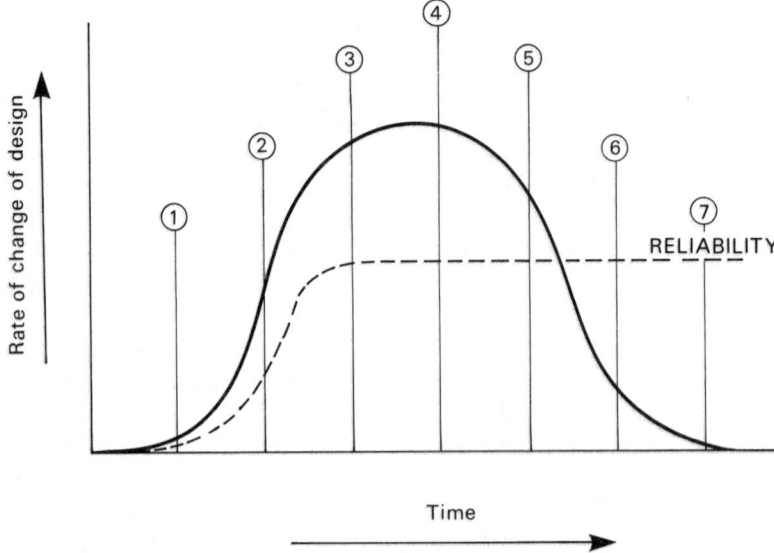

① Invention.
② Development.
③ Initial commercial introduction.
④ Technical Improvement.
⑤ Design adjustments under competitive economic pressures.
⑥ Application now at maximum profitability.
⑦ Replacement by new design.
　The dotted line curve represents the reliability of the product and reaches a peak where, without immediate redesign, reliability cannot be improved.

achieving a stimulus for the fostering of ideas. One such method is lateral thinking.

The systematic approach to design develops the designer's ability to solve a problem from the minimum of information. It sets out a method of applying various levels of design concept in order to build up a complete design, and can be prompted by the basic specification.

But suppose this is not available? During the inventive period, thoughts move around in the mind, deviating at times from the practical into the realms of fantasy. Under normal conditions these fantasy ideas are discarded without a second thought because they probably do not lead to an easily-derived solution. Also it is possible to become inhibited with all sorts of secondary features such as cost, size, form, aesthetic appeal, ergonomic features, etc. However, under the controlling influence of EXPERIENCE these more peculiar ideas can be made to give up the best or the real solution.

So what is Lateral Thinking? It is the ability to make the brain move consciously away from our normal OBJECTIVE route to a solution, and to think of other ideas, but ensuring that the branches of this alternative route never become totally disconnected from the main trunk and that the ideas can be guided back to the main objective route to the final solution. Fig. 1.9

Fig 1.9 Conceptive
design process

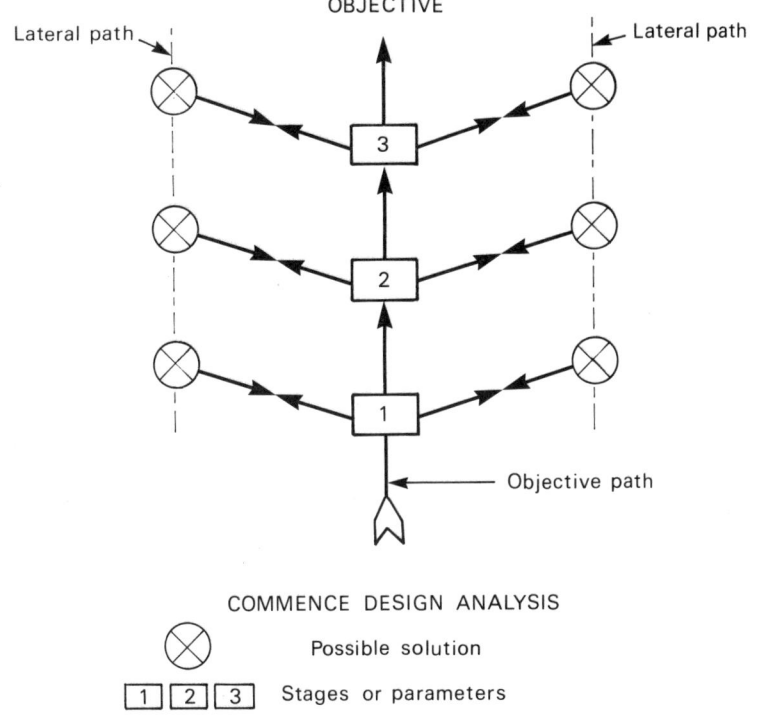

shows diagrammatically the interconnections of this conceptive design process.

The objective route which has numbered boxes along it is the quickest route to the solution of the problem. The numbered boxes represent the various stages to be evaluated. For example, suppose in the design of a machine that there are several items requiring individual design exercises such as mounting feet, tool post, and table drive. Fig. 1.10 shows that these items can be placed in the boxes to identify the individual design problems.

Taking them one at a time, the designer considers possible solutions. In the case of the table drive, ideas are listed in the lateral plane as shown in Fig. 1.11. Having listed possible solutions to the table drive, the designer then goes onto analyse the tool post, and this is done in the same manner as for the table drive. It should be noted that at no time does the designer become inhibited with thoughts about manufacturing methods, cost, aesthetics, etc.

The lateral route, indicated by the circle with a cross inserted, is intended to mentally disconnect the designer from the normal solutions, for example rack and pinion or bevel gears, and allows entry into the realms of fantasy, normally called an "excursion". Some of the ideas listed, of course, will appear quite reasonable at the end of the exercise, and it is only by a careful analysis of all solutions to all design problems encountered along the objective route that a real solution to the machine design can be realised.

Fig 1.10 The objective route

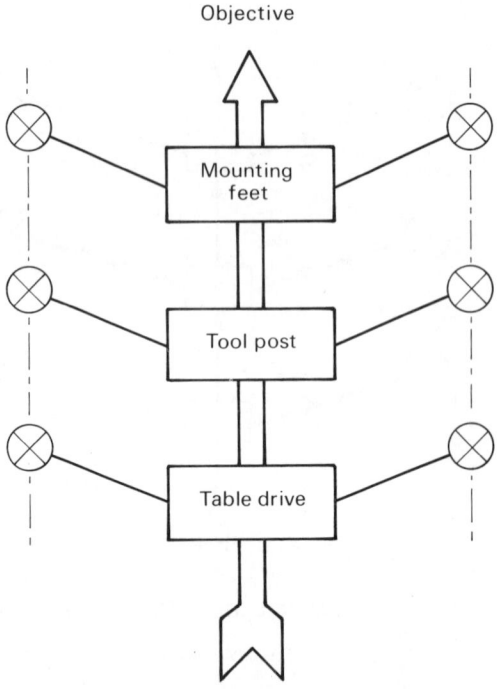

Fig 1.11 Listing of possible solutions to a design item
• = Excursions

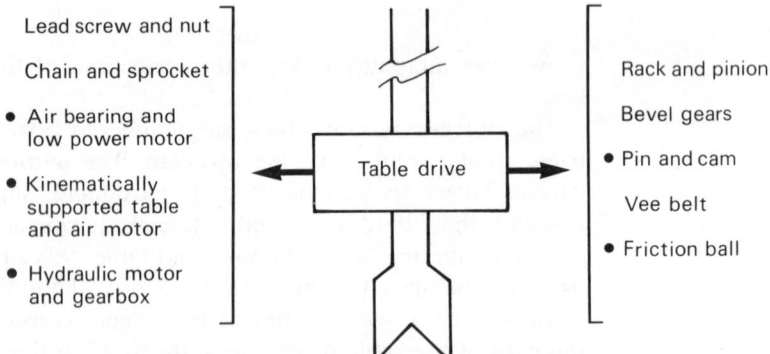

1.8 Group Stimulus

No design method gives perfect solutions every time, and a designer's approach can be unnecessarily restricted by emphasising a personal "feel" for the problem. Because of this, the process of lateral thinking is limited by each individual designer's attributes. The mind must have time to accustom itself to the environmental conditions that exist in the office or room where the designer is applying the thought process. Noise, for example, has an important influence on stimulus creation, and hence on problem-solving.

An individual designer's limited knowledge can in fact lead to a mental blockage and therefore prevent the solution to a problem. Hence, there is a school of thought that says a group of people with its diverse knowledge, perception and relationships with ideas can be dramatically more powerful in solving a problem than one person alone. However, relevant studies of performance by groups shows this to be dependent on the problem and mix of experience in the group.

It can be said that creativity arises when a goal is unachievable by normal reflex or habitual practises, and the first act is to recognise a problem and receive a stimulus, the second is to form a concept. Two such creative methods of problem-solving are Brainstorming and Synectics.

Brainstorming

Brainstorming is currently a popular term very often misused and misunderstood in engineering. It is commonly used just to mean "let's sit down and put our heads together to solve this problem". However, it is really a method whereby a group of engineers, designers, etc., get together in a session and throw ideas into a melting-pot. Ideas are amassed and these are sorted out after the brainstorming session.

A team or committee can very often be negative in its thoughts and very often this leads to mental roadblocks which prevent people with an idea wanting to contribute to the problem. Care should be taken to recognise roadblocks and eliminate them. Typical roadblock expressions include:

"That will not work".
"Experience has shown that it will not work".
"It is against company policy".
"We have tried that before".
"Adhesives won't work".
"Manufacturing costs will be too high".

The team selected to perform the task of problem-solving should have a wide range of experience, and preferably be from various departments. For example the team may consist of designers, production engineers, sales engineers, and a marketing expert.

As ideas are thought of by team members, they are written down by a person assigned this task. The aim should not be to sort through and evaluate the ideas at this point of the proceedings; and because of the natural tendency for people to want to extend their ideas, care should be taken to ensure that this does not occur. Consider each of the items in turn on their merit, moving onto the next one only when it is considered that the ideas have been exhausted.

The problem with this particular method of problem-solving is the amount of noise that can be generated due to lack of leadership. The more extrovert members having strong ideas can turn the session into chaos, resulting in negative results and roadblocks.

Synectics

Companies who look for ways of reducing their costs and increasing their profitability may turn to the process of Synectics. This is a group problem-solving method, but unlike brainstorming depends on a leader to guide the process through to a satisfactory solution. The method is as follows.

a) A *leader* is chosen to steer a group of people from various company departments in a prescribed way, but may not contribute to the business of the group.

b) Several groups from the company, made up of varying skills, are given time to select a solution.

c) The chief role is that of the *client* with a problem. The solution can be implemented at the end, because the *client* has sole authority to implement it.

d) There is an *observer* who is trained in synectics and watches the session, and who, afterwards, leads a discussion replaying parts of the best three ideas or points, and the worst one.

e) At the outset the *client* describes the problem, indicates the measures already taken, and suggests the ones with which help is expected. Questions are unnecessary; they are in any case discouraged and regarded as a form of noise. During this period *participants* write down their ideas.

f) Concepts are then suggested which, if realisable, might lead to a solution. These are written on a blackboard by the *leader*.

g) To ensure a positive atmosphere, concepts are expressed as a "wish" and participants are trained to start with the words "How to ...." Since a large quantity of ideas emerge at this stage, possibly in a stream, it becomes extremely difficult to identify their source. Also at this stage, no evaluation takes place, therefore no criticism, hence a feeling of goodwill exists.

h) Ideas end when the *client* selects one that is acceptable and explains why it is.

i) The *leader* then seeks one other of the ideas from the *participants* which must fulfil the expressed "wish" of the *client*. To ensure that the client understands this idea, the *client* repeats the idea and then gives three reasons why it is a good one and outlines one reservation, should there be one. The *leader* then calls for an idea to meet the expressed objection and the process is repeated until the *client* has no further reservations.

j) At the *leader's* discretion, the *participants* can be invited to expand an idea before the *client* expresses opinion. This procedure may rapidly produce a solution, although one is not agreed until the *client* declares that it now satisfies requirements.

Table 1.3 shows a typical chart of idea response obtained from ten sessions. It should be noted that the more prolific members, e.g. *participant* (C) in session 10, do not necessarily have their ideas accepted; in this case the asterisk against (B) indicates that one of his ideas was ultimately accepted as a solution.

Synectics is not an easy method to adopt because of the extreme pressure on the *client* to accept a solution. The client often ends up exhausted at the end of a session. However, because of the organised way in which the method is carried out, a solution is always possible.

Table 1.3 Distribution of ideas

Participant	Session no.										Average no. per Participant
	1	2	3	4	5	6	7	8	9	10	
A				6*	11		2	6	7	8	6.7
B		3				1				8*	4.3
C		6*		6	9	11				14	9.2
D			3	4	5*	8*	2	5		5	4.6
E	4								13	9	8.6
F	2	1								3	2.0
G				3	1					1	1.6
H		2							7		4.5
I									4*		4.0
Total	6	12	3	19	26	20	4	11	31	48	

* Idea adopted from these.

**1.9
Value Analysis**

Value analysis and value engineering are concerned with ensuring that the designer is conscious of the ways in which production methods and form design can influence the economics of producing a component.

The growth of these two subjects arose out of the need to find substitute materials during the second world war. In some cases it was necessary to revise designs to enable these substitute materials to be used. This sometimes resulted in a better product which cost less.

Today, value engineering is concerned with finding the best solution at the drawing board stage, and it should be the aim of all designers. There is, however, the case where a product may be losing money and value analysis is used to look into ways of cheapening the product.

Value analysis starts by looking at the function of the component. Alternatives are developed and analysed by an organised team which derives the means of performing the function at the lowest cost with the appropriate quality.

**1.10
Innovative Design**

As stated earlier in this chapter, mental roadblocks can occur which prevent the solution to a design problem. These normally occur through the designer's inability to become divorced from old habits and familiar modes of thought, developed over many years of inadequate design practices.

To be an innovative designer it is necessary to divorce the mind from the practicalities of manufacturing techniques, the life of the product, and particular types of material. The sole interest is in meeting the functional requirements of the design.

This process means that the cost of producing the part is not a priority in the designer's mind. This does, of course, defeat the object of value engineering, but cost savings can be achieved once the design strategy has been resolved, and this process is one that must be entered into knowingly.

What is innovation? When applied to design it is the ability to think of a new way of doing something which has always been done by conventional or habitual means. For example, take the simple tin opener which for years had been produced as a levered cutting edge (Fig. 1.12a). We now have the gear cog tin opener which uses the edge of the tin as a kind of railway track to cut

Fig 1.12 Evolution of the hand-held tin opener
[c William Levene Ltd.]

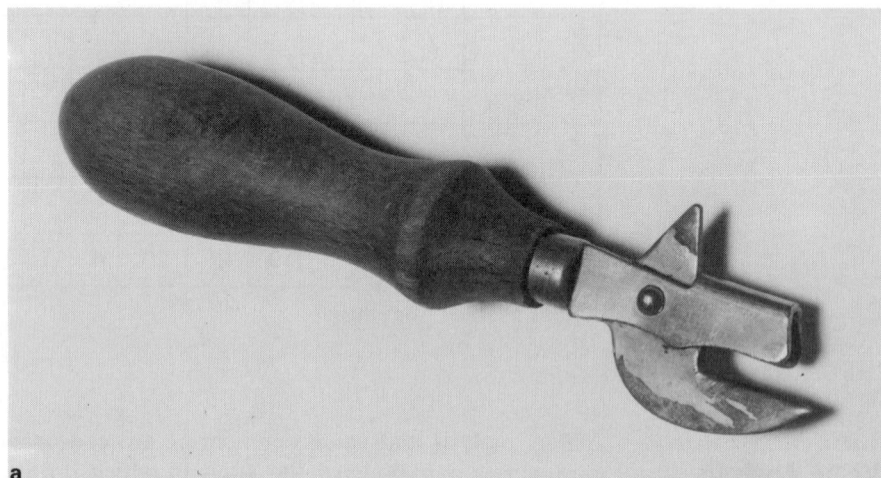

a

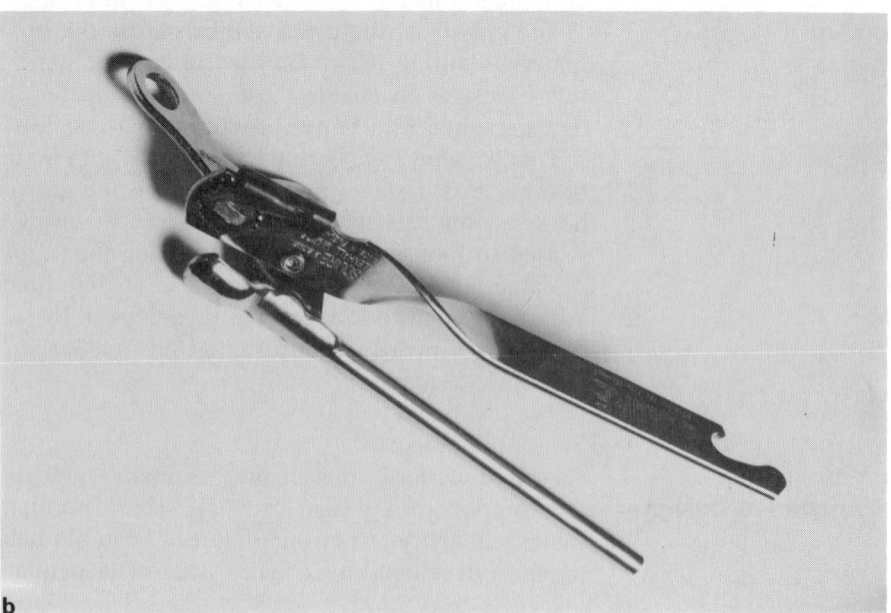

b

through the top of the tin (Fig. 1.12b). This device was a breakthrough as a design because suddenly the effort required to open tins was reduced and the simplicity of the design made sure that it was extremely cheap to make. The plastic "Lift Off Can Opener" shown in Fig. 1.12c marks a further stage of innovation in the design of hand-held openers. The main principle of this device is the split handle which pivots the cutting wheel and disc about each other on an eccentric. Thus they are farthest apart when the handle halves are fully open, and are connected for cutting when the halves are closed. Innovative features include:

a) A clean safe cut around the *side* of the can.
b) Little effort to rotate cutter, and *no* effort to keep handle halves closed.
c) Completely new shape with mainly plastic construction.

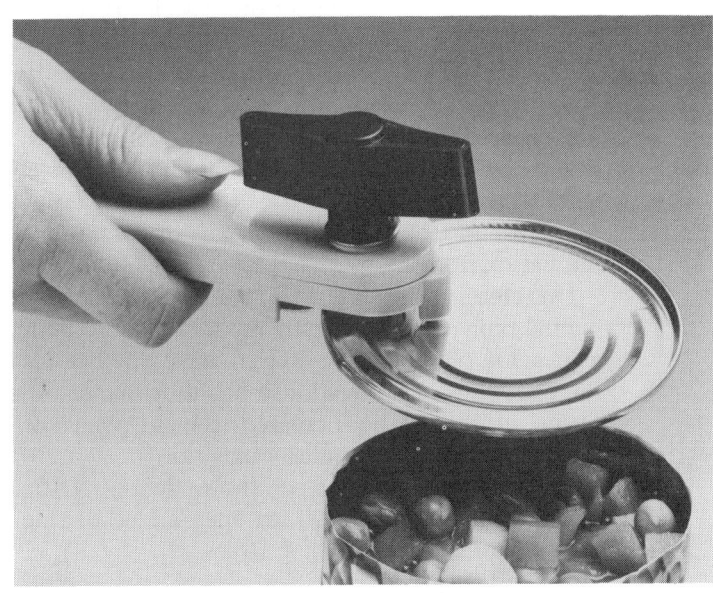

Another example is the adaptation of the transverse engine to motor vehicles. Alec Issignosis' idea as applied to the Mini was a breakthrough in new technology (Fig. 1.13). It meant that there was no longer a need to supply a rear drive axle to a car and hence it eliminated the need for expensive half shafts, hypoid gears, etc. It also allowed the use of a flat pan within the passenger compartment.

The procedure needed to produce an innovative design cannot be set out by way of a set of rules; only by getting away from conventional solutions can an innovative design be produced.

Fig 1.13 Sectional view of the original Austin-Morris Mini [British Leyland]

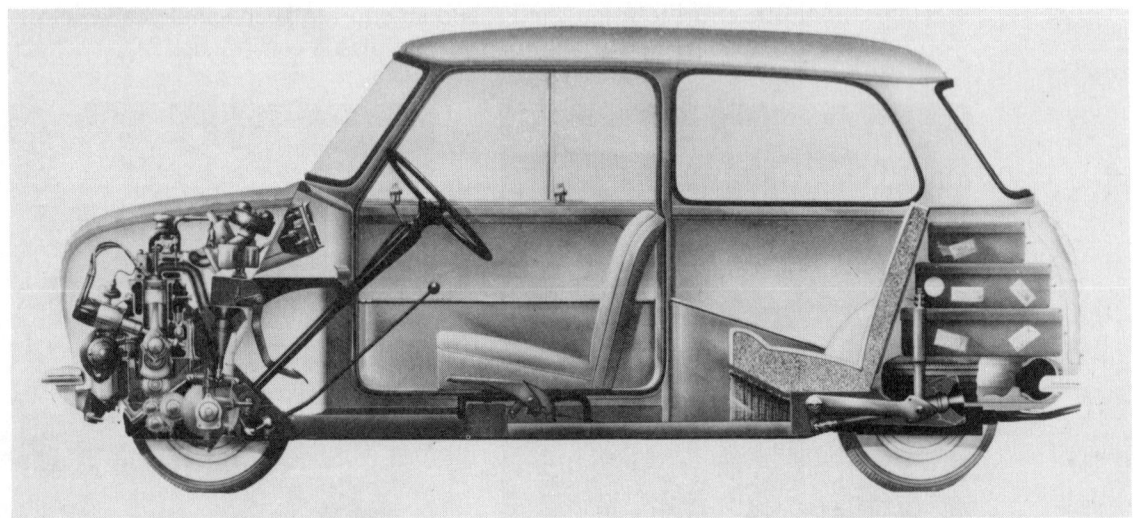

1.11 Design Innovation—A Series of Case Studies

The following recent case studies are all good examples of innovative flare as an essential ingredient in the design process.

1. "Flexiwrench" Multi-Angle Tool System

This remarkably universal tool was invented by Mr. W. J. Gornal and developed in conjunction with staff and students at Preston Polytechnic. It is an obstruction spanner whose tool head may be swivelled into a pre-selected position and locked rigid by means of a cam-lever handle. Fig. 1.14 shows the basic arrangement.

On releasing the slide-lock, the cam-lever may be pivoted, thus disengaging the plunger and allowing the tool head to be moved through a variety of angles. On obtaining the desired angle, the head is again locked by reversing the procedure. The head also has four planes of rotation at 90° in its axial plane, which may be increased to eight planes at 45° by using an optional bi-square drive with a plastic adaptor.

With the combination set shown in Fig. 1.15, over 100 000 alternative positions may be achieved for open-ended or ring tool heads. Some combinations are shown in Fig. 1.16.

Fig 1.14

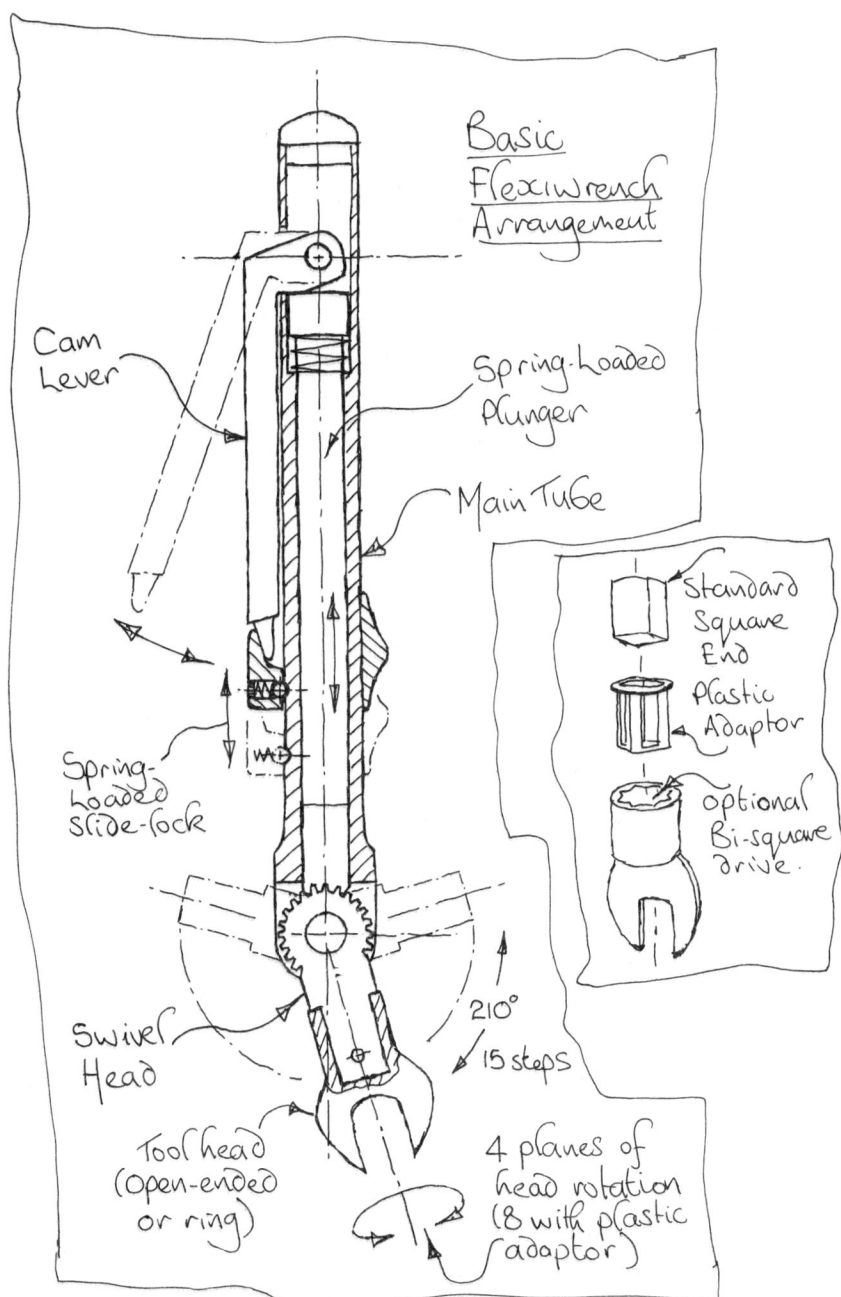

Basic
Flexiwrench
Arrangement

Cam
Lever

Spring-Loaded
Plunger

Main Tube

Spring-
Loaded
Slide-lock

Standard
Square
End

Plastic
Adaptor

optional
Bi-square
drive.

Swivel
Head

210°

15 steps

Tool head
(Open-ended
or ring)

4 planes of
head rotation
(8 with plastic
adaptor)

Fig 1.15

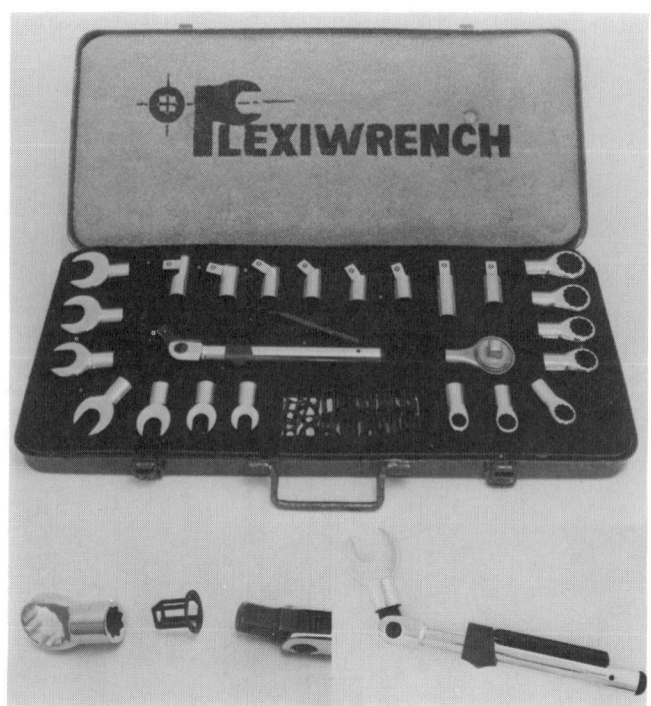

Fig 1.16

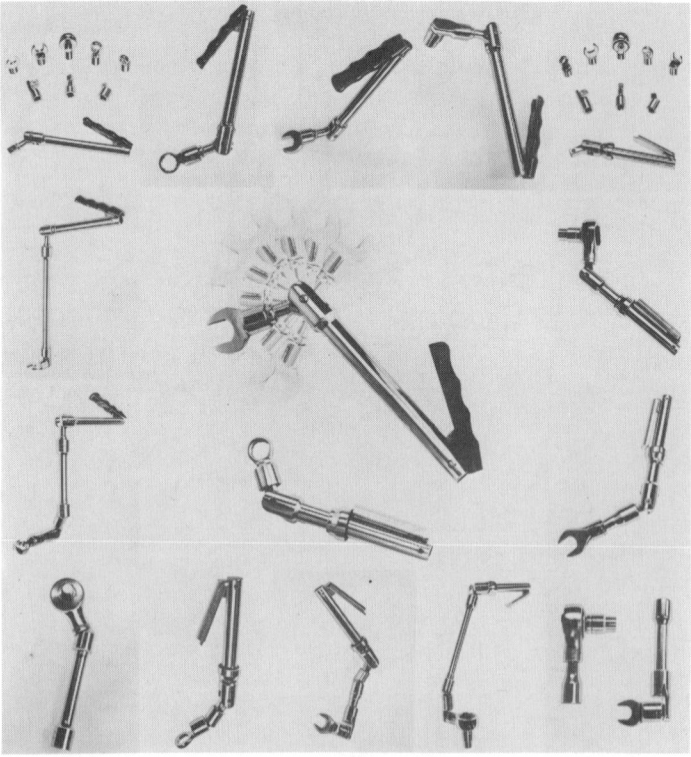

2. Centrifuge for Medical Analysis (Microprocessor-controlled)

The "Cytospin 2", shown in Fig. 1.17, is a programmable centrifuge, designed by Shandon Southern Products Ltd., aided with a grant from M.A.P. (the Department of Industry's programme to encourage the application of microelectronics).

The innovation in this design enables medical analysts to obtain microscope slides of human body cells quickly and accurately via separation from a small sample of fluid. The blank slide is attached to a filter card and fitted, with the sample chamber, into a slide clip as shown in Fig. 1.18.

The assembled clip is then put into its place in the centrifuge head, the chamber filled with the fluid sample, and the sealed-head fitted to a taper on the drive motor shaft.

Fig. 1.19 shows the basic principles of operation. During fast rotation of the head, centrifugal forces cause the sample to pass through a hole in the filter card. The cells, being denser, are more centrally biased and thus reach the slide behind the card, whilst the lighter fluid is forced out and absorbed by the card. The required time, speed, and acceleration value is programmed and controlled via a microprocessor. Accuracy is monitored by counting light pulses from an L.E.D. (light-emitting diode) through a slotted disc on the motor shaft. These are received by a photo-diode which sends an electrical feedback signal to the microprocessor. This compares the feedback signal with the required quantity and continues to send a correction signal to the motor until the programmed command value is achieved (Fig. 1.20).

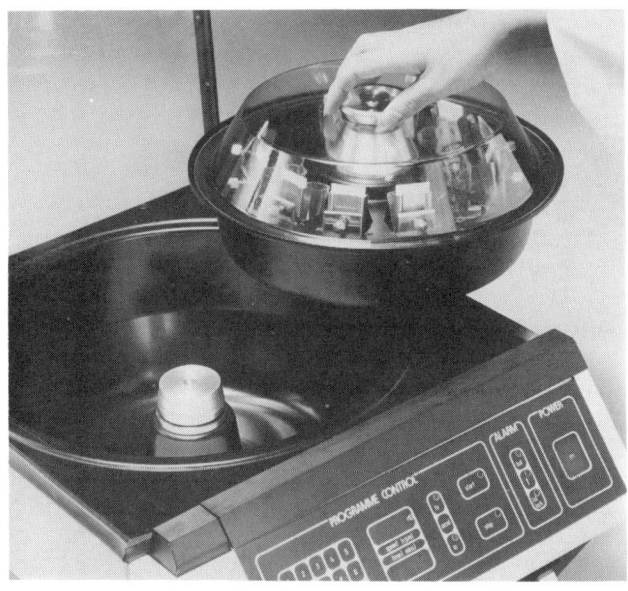

Fig 1.17

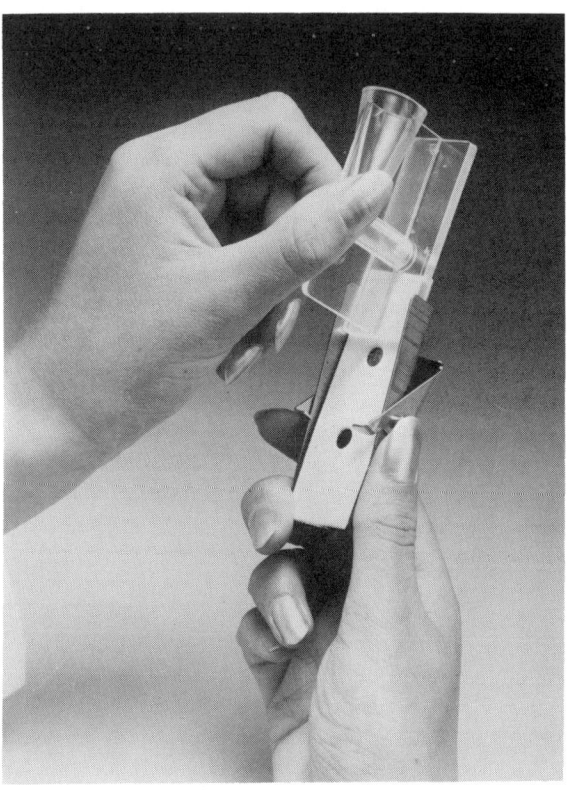

Fig 1.18

Fig 1.19

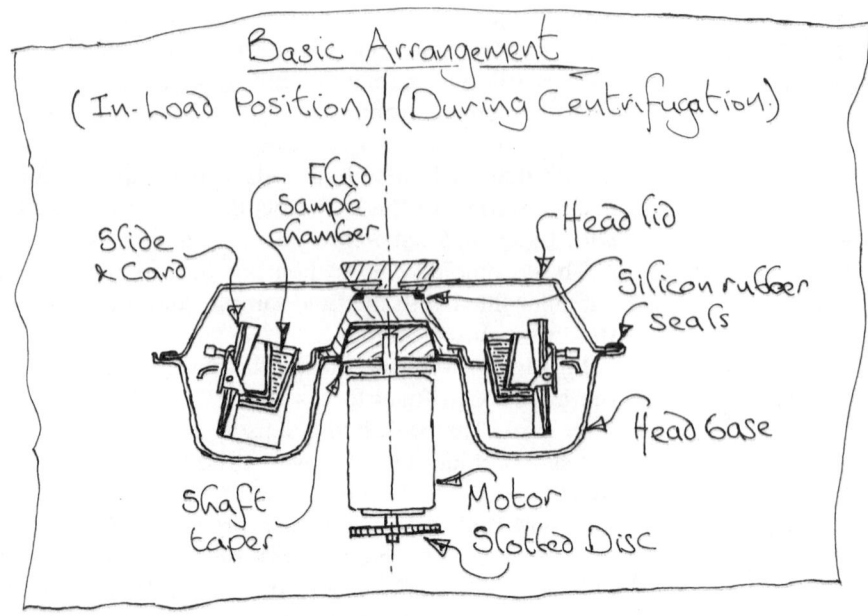

Basic Arrangement
(In-Load Position) | (During Centrifugation)

Fluid Sample Chamber
Slide & Card
Head Lid
Silicon rubber seals
Head Base
Shaft taper
Motor Slotted Disc

Fig 1.20

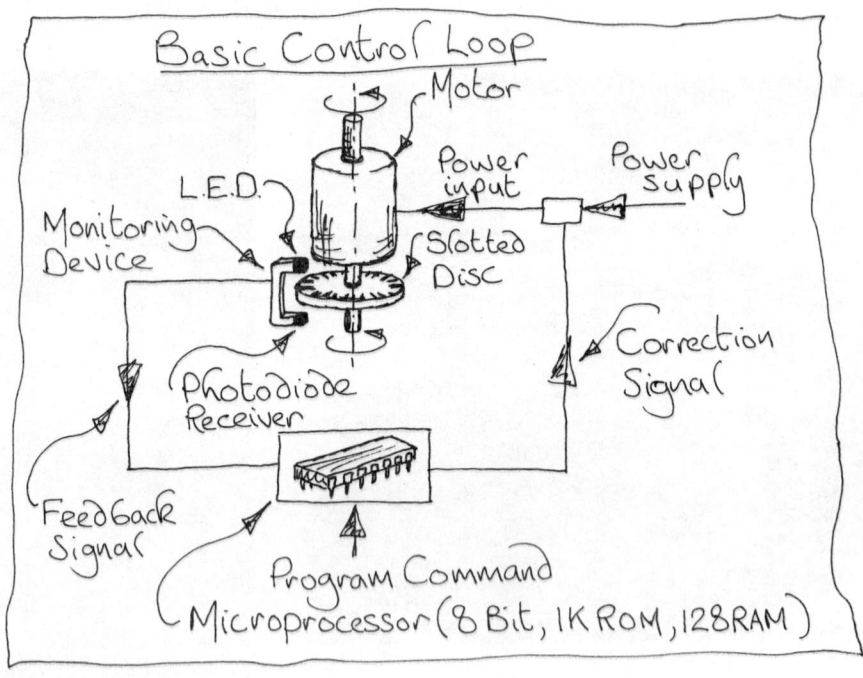

Basic Control Loop

Motor
Power input
Power Supply
L.E.D.
Monitoring Device
Slotted Disc
Correction Signal
Photodiode Receiver
Feedback Signal
Program Command
Microprocessor (8 Bit, 1K ROM, 128 RAM)

3. Design for Compactability

Changing social trends such as smaller houses and cars; increased and diversified forms of travel; and an upsurge in D.I.Y. enthusiasm, have created the demand for a whole range of product designs which are as small and compact as possible when storing and transporting, as light as possible when handling and transporting, and yet maintain efficiency of function.

One consequence has been a growing list of innovations in light-weight folding equipment, such as bicycles, children's pushchairs, invalid's wheelchairs, D.I.Y. workbenches, and camping accessories.

Design for compactability has two essential requirements:

a) Full exploitation of modern light-weight materials such as aluminium tubing and plastics.

b) Creative flare in devising efficient compacting methods which are suitable for these component materials and their methods of manufacture.

The "Globetrotter Travel Iron" shown in Fig. 1.21 (courtesy G.S. Iona Ltd.) is an excellent example of innovative design form to suit compactability.

Fig 1.21

4. Toy Design

Toy design offers plenty of scope for good innovation, particularly in the areas of ingenious mechanisms, electronic circuitry, and aesthetic creativity.

Toys must, of course, be designed to stringent safety requirements, which are broadly listed in BS 5665(1979) *Safety of Toys*. Obvious considerations include the use of non-poisonous and non-flammable materials; avoidance of sharp edges; and adherence to the appropriate product and component sizes for a recommended age range (small components, for example, may be easily swallowed by very young children).

Common types of plastics moulding can generally conform to these requirements and are thus extensively used. They have the added advantage of being available in a vast range of attractive colours and thus do not need to be painted.

The innovation of the Fisher Price Toys shown in Fig. 1.22 lies in their aesthetic novelty and imaginative educational features. For example, the "Terry Tape Measure", shown centre, is highly original and effective as a first course in measurements for a young child.

Fig 1.22

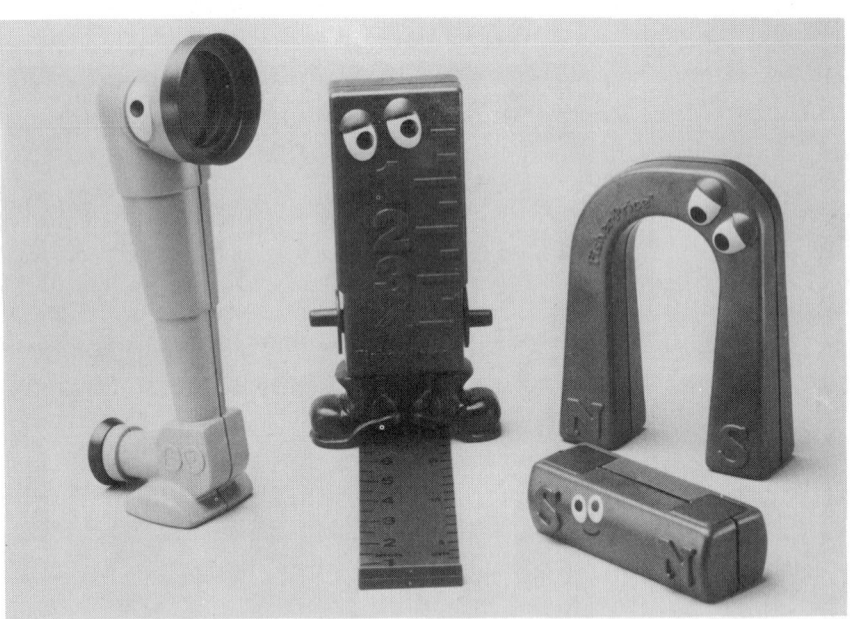

5. High Technology Artificial Limb

Fig. 1.23 (courtesy Chas. A. Blatchford & Sons Ltd.) shows the "Endolite" Prosthetic system—almost certainly the world's most advanced artificial lower limb.

This truly second generation system contains miniaturised functional elements in order to "package" the limb in a full-length polyurethane foam cosmetic fairing and silicone rubber outer skin. Its basic working principles are shown in Fig. 1.24.

The main areas of innovation are as follows:

a) Extensive use of high technology lightweight materials including carbon fibre composites for the structural elements, advanced polymers and elastomers in the flexible items, and titanium alloy fixings.

b) The "Multiflex" foot/ankle mechanism which incorporates a rubber ball joint and snubber ring arrangement providing maximum freedom of ankle movement. Toe flexion is provided via a polyester elastomer strip attached to the carbon fibre foot base. A serrated fixing provides heel-height adjustment.

c) A knee mechanism which swivels against a pneumatic damper until weight is applied, when an automatic brake is activated to lock the knee joint in position. This allows the user to maintain the necessary rigidity during stance position (when weight is applied) and the necessary free movement during flexion position (when weight is not applied), as shown in Fig. 1.25.

A spring mechanism releases the brake when the weight is retracted. Inward movement of the knee-cap is provided via a spring-loaded cam when bending, thus avoiding stretching and damaging the outer skin.

As with all such equipment produced in the U.K., this design has been fully tested to conform to the rigorous safety standards specified by the Department of Health and Social Security Scientific and Technical Branch.

Fig 1.23

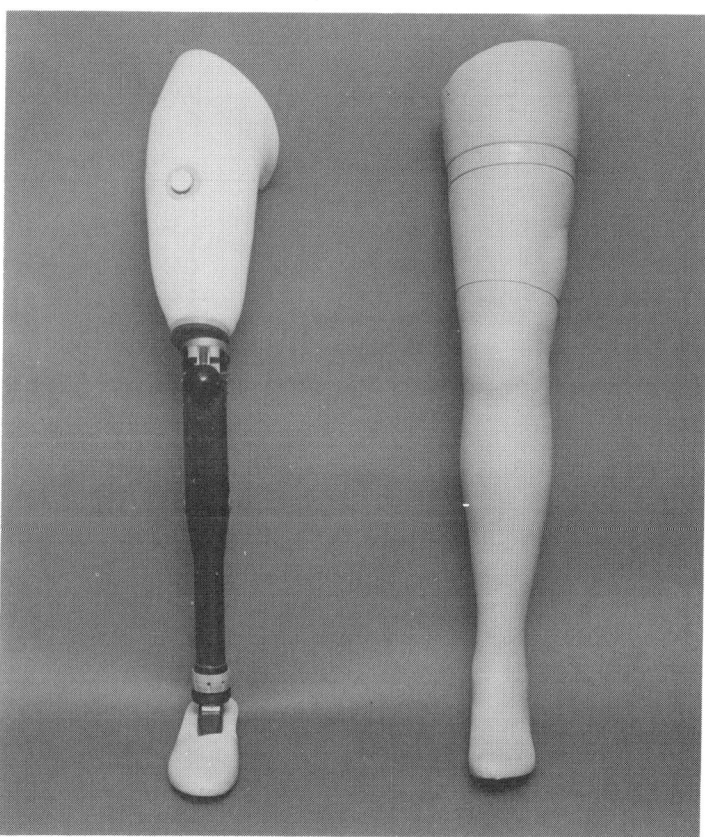

Fig 1.24

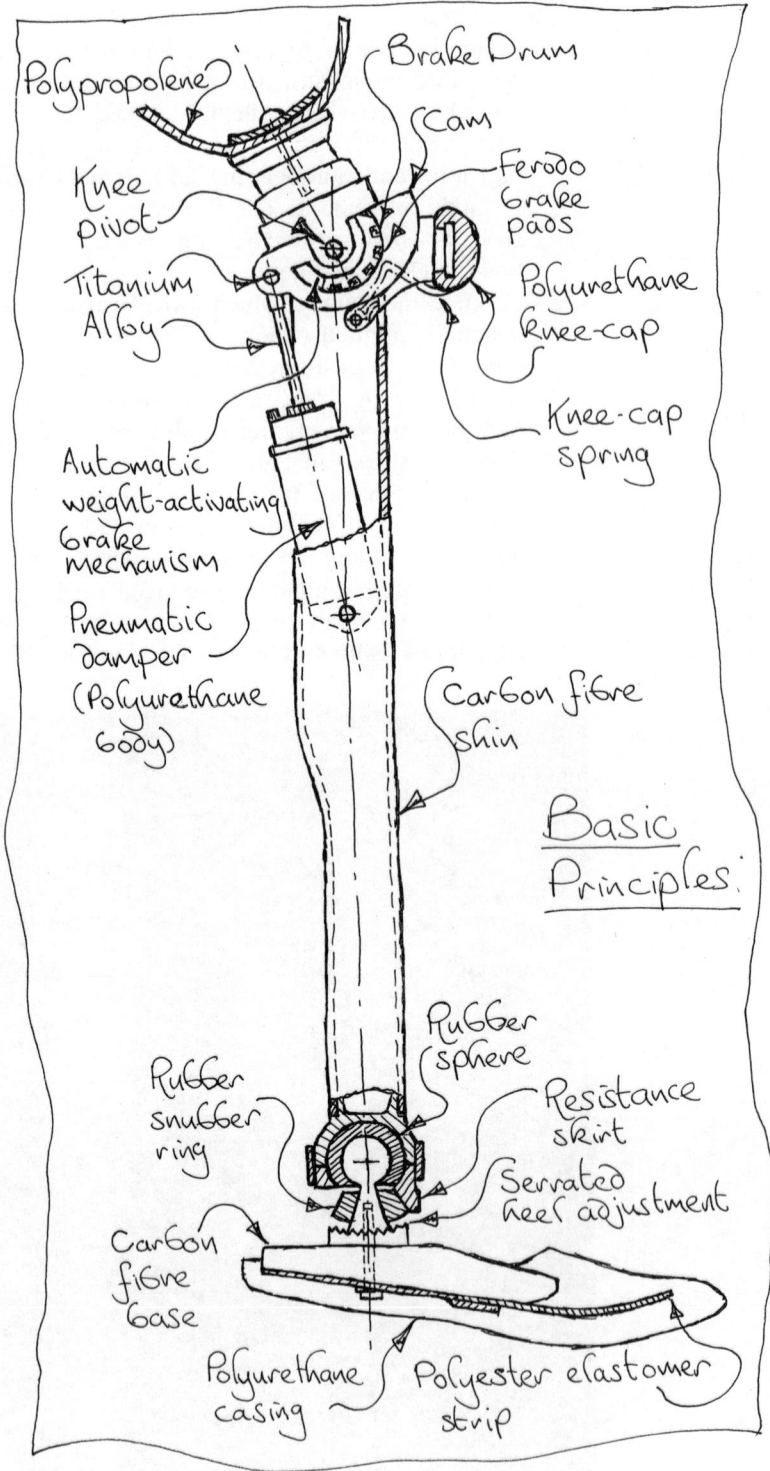

Polypropolene

Brake Drum

Cam

Knee pivot

Ferodo brake pads

Titanium Alloy

Polyurethane knee-cap

Knee-cap spring

Automatic weight-activating brake mechanism

Pneumatic damper (Polyurethane body)

Carbon fibre shin

Basic Principles:

Rubber snubber ring

Rubber sphere

Resistance skirt

Serrated heel adjustment

Carbon fibre base

Polyurethane casing

Polyester elastomer strip

Fig. 1.25

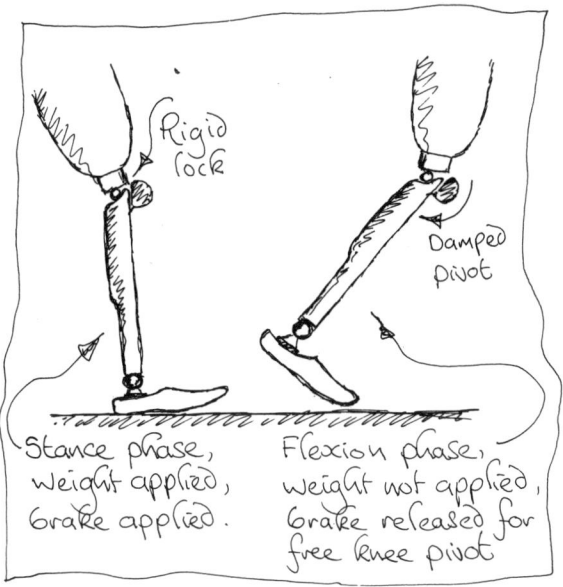

Stance phase,
weight applied,
brake applied.

Flexion phase,
weight not applied,
brake released for
free knee pivot

6. Self-Elevating Vehicle Lift

Fig. 1.26 shows an entirely new concept in raising vehicles for repair.

The simple "Hy-Ramps", produced by Brockhouse Harvey Frost Ltd., employ a see-saw action and operate using hydraulically-damped tracks balanced on pivots. On entry, the ramps are inclined as shown in Fig. 1.27.

When the vehicle passes over the balanced point, the dampers gently lower the ramps to the horizontal position giving a ground clearance of over 0.7 metres, as shown in Fig. 1.28.

The combination of extra ground-clearance, safety, and portability, make Hy-Ramps attractive to garages, fleet users and D.I.Y. enthusiasts—particularly for lengthier and on-site repairs (Fig. 1.29).

Fig 1.26

Fig. 1.27

Fig 1.28

Fig 1.29

7. Intermediate Technology for Indian Villages

Prohibitive costs, and difficulty in accounting for appropriate local needs, often make modern technology unsuitable for inhabitants of the poorer developing countries, particularly those in the remote villages.

Intermediate, or appropriate, technology, is thus evolving as a cheaper, practical alternative in these areas. Most intermediate technology designs rely on the principle of "ad-hocism", which involves adapting resources immediately at hand rather than conforming to rigid design schemes.

Bicycles are used everywhere in India and are thus an obvious choice for intermediate technology in that country. Fig. 1.30 shows five simple design adaptations using bicycle parts as described in the *Engineering Designer* magazine by Mr. C. B. Suresh Babu.

Arrangement (a) shows a manually-operated eye-lens surface grinder. The lens is fixed in a small chuck which is rotated by pedalling. Grinding paste is applied to a concave tool and pressed against the lens via a handle.

Arrangement (b) shows a pedal-powered pump used in villages where electricity is not available or by marginal farmers who cannot afford power-driven pumps.

Arrangement (c) shows a wheelbarrow in which the large wheels reduce rolling friction and aid transport of farm produce through sand or slush.

Arrangement (d) shows a bicycle side carrier which provides a cheap, stable load-carrying device.

Arrangement (e) shows a two-in-one concept for people who are orthopaedically handicapped below the waist. When the castor is fixed, it serves as a wheelchair for mobility in the house, and when the hand-driven front wheel is fixed, it becomes a tri-wheeler for outside transport.

Fig 1.30

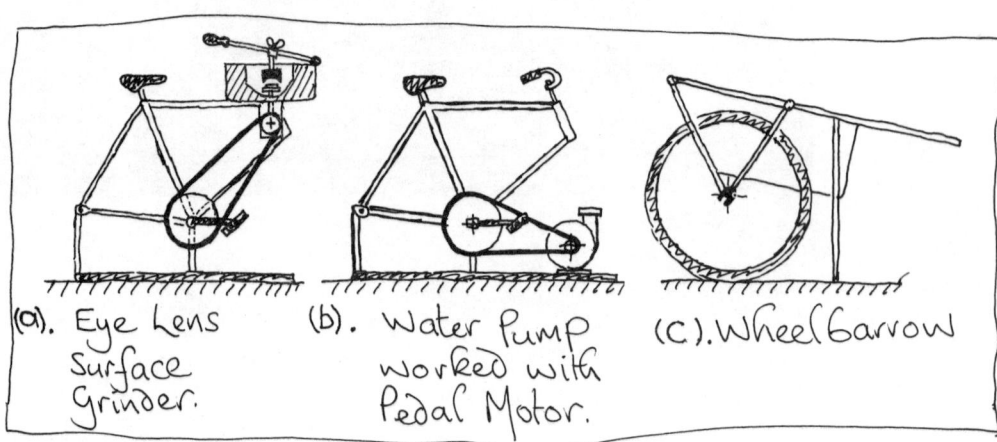

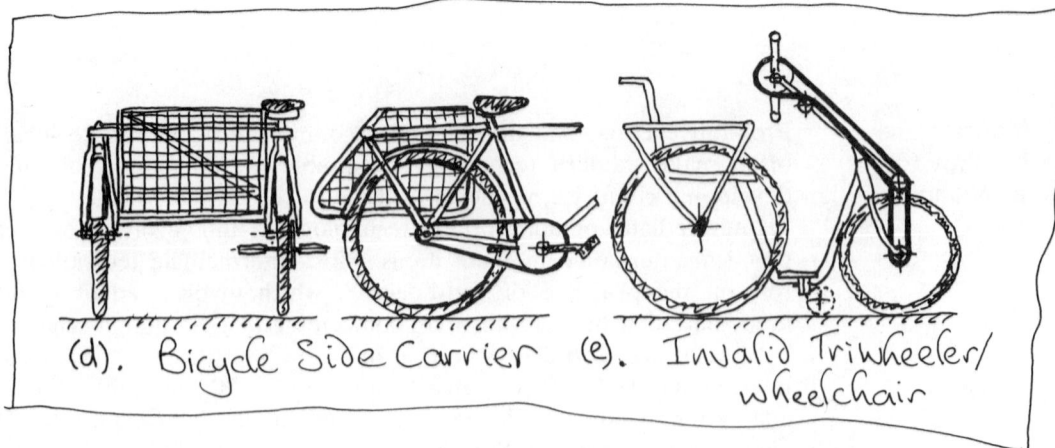

Problems

1.1 In the design process it is necessary to follow a systematic route for choosing the optimum design. Fig. 1.1 shows a route which could be taken. Briefly explain the actions required of the designer at each stage of this route. Give reasons for taking these actions, and outline any input required from outside sources, such as the customer.

Fig 1.31

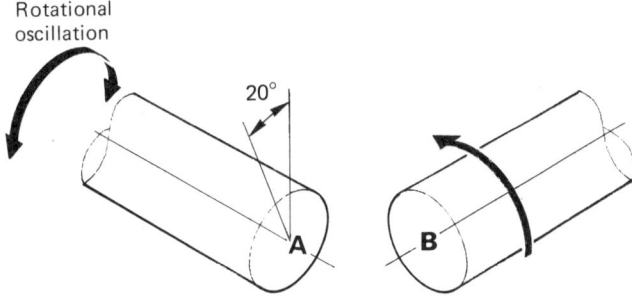

Rotational oscillation

20°

A

B

1.2 Fig. 1.31 shows two shafts positioned at right angles to each other, in the same plane. This is part of a machine drive and it is required to design a mechanism such that shaft A will oscillate 20° about the centreline, as shown, while shaft B is continuously rotating. Carry out a systematic design evaluation, preparing sketches, and analyse each proposal to find the best solution. Construct tables similar to 1.1 and 1.2 to assist your evaluation process.

1.3 Prepare a specification for the mechanism designed in **1.2**. Give details of the type of machine it is to be used on, using a hypothetical example. Follow the rules given in section 1.3.

1.4 As stated in section 1.6, designs go through life cycles. Using the example of a bicycle, write a report based on Fig. 1.8 outlining the activities occurring at each stage from 1 to 7.

1.5 What in your own words do you understand by the term Lateral Thinking? Use an example to illustrate your understanding of the subject.

1.6 Group stimulus is used as a method of arriving at design solutions. What methods are available to assist in this process? Write a report using your own words to describe a typical Brainstorming session.

1.7 Attempt to produce an innovative design of a table such that, even if placed on an uneven floor, the top always remains level.

1.8 Produce sketches showing innovative designs of a car jack. Using the systematic design process as in Tables 1.1 and 1.2, evaluate to find the best design solution.

2 Design Planning Methods and Organisation

An efficient design procedure is unlikely to be achieved without good organisation and scheduling. In order to ensure that a design project is completed on time, the designer must take steps to see that a suitable plan is prepared. This plan must take account of such things as material selection, detail design detailing, checking, etc. Two common techniques used in project planning are

1) Gantt or bar chart.
2) Network planning charts.

2.1 The Gantt Chart

This was the first formal method used to establish start and finishing times for various phases of a project. It was devised by Henry L. Gantt, an adherent of Frederick W. Taylor, at the turn of the twentieth century. It is still widely used today because of its simple style of presentation.

Generally, a time scale is placed horizontally along the top of the Gantt chart. The rows represent the activity to be scheduled. Fig. 2.1 shows a typical Gantt chart, for planning the design, ordering of parts, and building of a hydraulic motor.

It should be noticed that some activities can be carried out whilst other related ones are also going on. For example, the checking of completed details can be done whilst the remaining detailing is being completed.

2.2 Network Planning

This technique lends itself to more complex activities and is thus used more frequently than Gantt charts when large projects are being planned. One particular advantage of network planning is the priority rating for each activity in the project.

Low-priority activities are those which can be "shelved" without affecting the overall completion date.

High-priority activities cannot be delayed without extending the project duration. High-priority activities are said to lie on a *critical path*. The process of representing a critical path in diagram form is termed **network analysis**. A network consists of the following elements:

a) *Activities*—these are denoted by straight lines. The description of each activity is usually placed above the line; and activity time allowance may be placed below.

b) *Completions–of–activities*—these are denoted by circles which include: an activity number; the earliest possible completion time; and the latest possible completion time (Fig. 2.2).

Fig 2.1 A Gantt chart

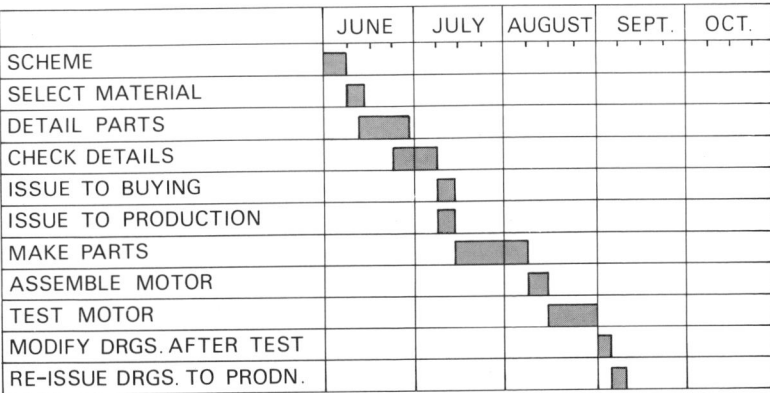

	JUNE	JULY	AUGUST	SEPT.	OCT.
SCHEME					
SELECT MATERIAL					
DETAIL PARTS					
CHECK DETAILS					
ISSUE TO BUYING					
ISSUE TO PRODUCTION					
MAKE PARTS					
ASSEMBLE MOTOR					
TEST MOTOR					
MODIFY DRGS. AFTER TEST					
RE-ISSUE DRGS. TO PRODN.					

Fig 2.2

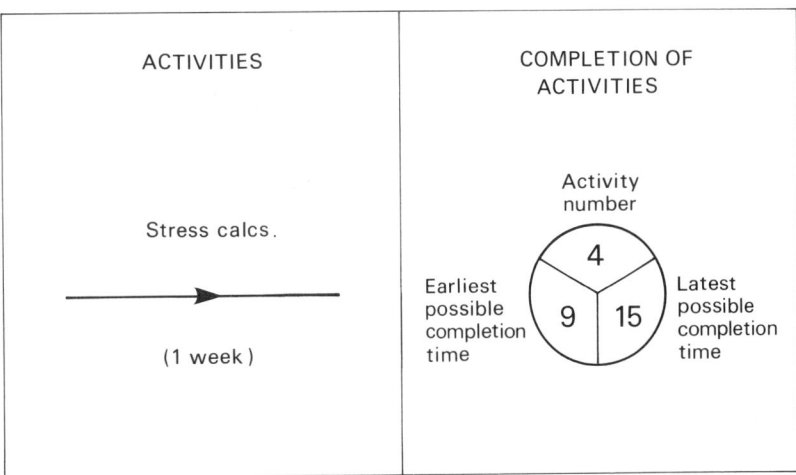

In order to illustrate how to construct a network, compare the simple task of making unbuttered toast, with that of making buttered toast. As shown in Fig. 2.3, a list of activities and their time durations must first be compiled. Activities which are contiguous on each other are then connected via their respective lines and circles. Earliest completion times may then be inserted by making progressive additions of the time durations. Latest completion times are then inserted by subtracting back through the network.

1) In the case of unbuttered toast, the network has only one route. No activity can begin until the previous one has finished. Earliest and latest completion times for each activity are thus the same.

2) In the case of the buttered toast, activities 1, 2 and 3 are not contiguous and can thus radiate from the same starting point. Where several activities converge onto one completion circle, it is the highest value of earliest completion time which is inserted.

Fig 2.3a Making
unbuttered toast

Fig 2.3a Making unbuttered toast

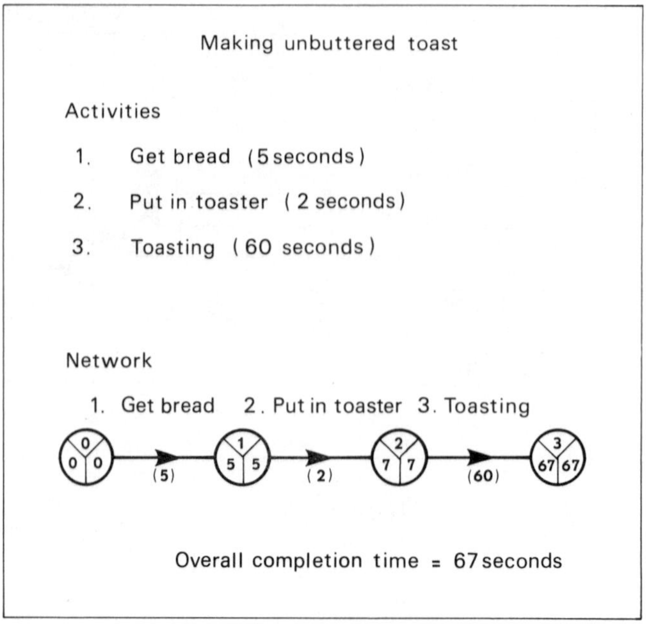

The critical path may be recognised as the line with the equal earliest and latest completion times, i.e.

Critical path = line $0 \rightarrow 1 \rightarrow 4 \rightarrow 5 \rightarrow 6$

Thus these are all high-priority activities and any delay in these will affect the overall completion time. The critical path also determines the value of the overall completion time, which is inserted in the final circle.

Tasks 2 and 3 are low-priority activities and may be delayed without affecting overall completion.

"*Dummy activities*" are those in which no actual process takes place but involve a situation where a new activity cannot be started until another one has been completed elsewhere in the network. These take no time. For example, activity 6 could not commence until activities 2, 3 and 5 have been completed. Dummy lines thus converge onto activity 5 completion circle on the critical path.

Latest possible completion time for task 2

$= 67 - 0$

$= 67 \text{ s}$

Latest starting time $= 67 - 5 = 62 \text{ s}$

Latest possible completion time for task 3

$= 67 - 0$

$= 67 \text{ s}$

Latest starting time $= 67 - 3 = 64 \text{ s}$

Fig 2.3b Making
buttered toast

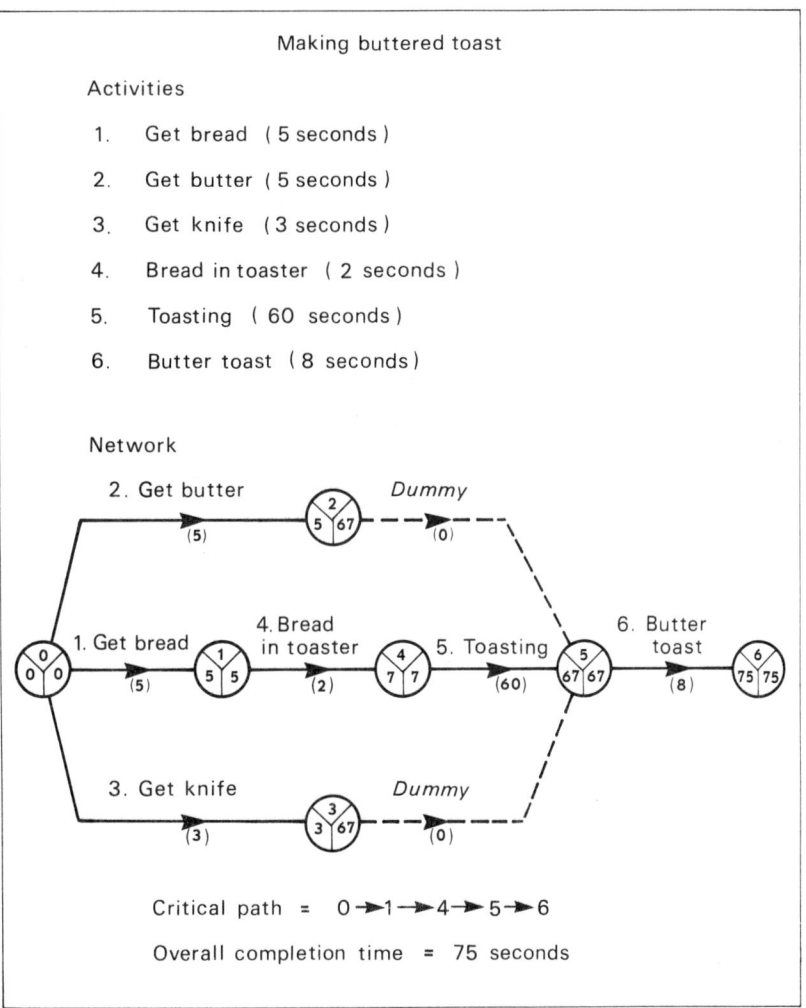

Making buttered toast

Activities

1. Get bread (5 seconds)

2. Get butter (5 seconds)

3. Get knife (3 seconds)

4. Bread in toaster (2 seconds)

5. Toasting (60 seconds)

6. Butter toast (8 seconds)

Network

Critical path = 0 ➤ 1 ➤ 4 ➤ 5 ➤ 6

Overall completion time = 75 seconds

We will now apply network analysis to the design and manufacture of a simple transmission assembly, shown in Fig. 2.4. The product is assembled and sold by a company which makes the fabricated steel items and buys out the remaining items. Fig. 2.5 shows the list of activities and the completed network. The critical path will be recognised as line

$$0 \rightarrow 1 \rightarrow 7 \rightarrow 8 \rightarrow 9 \rightarrow 10$$

and thus the overall completion time as 8 weeks.

It will be noted that the critical path involves the ordering and delivery of a bought-out item. Lengthy delivery times often affect overall completion times and these items should therefore be ordered at the earliest possible stage (i.e. as soon as selection calculations are complete, in this case).

Fig 2.4 A simple transmission assembly

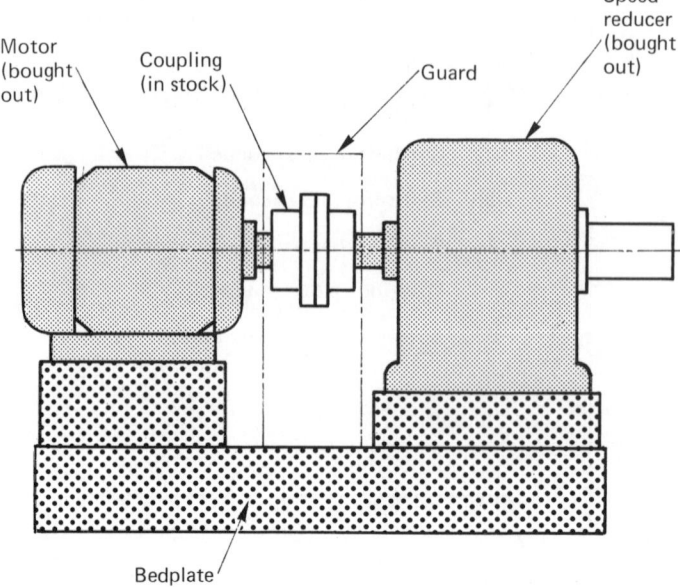

Fig 2.5 Activities and completed network

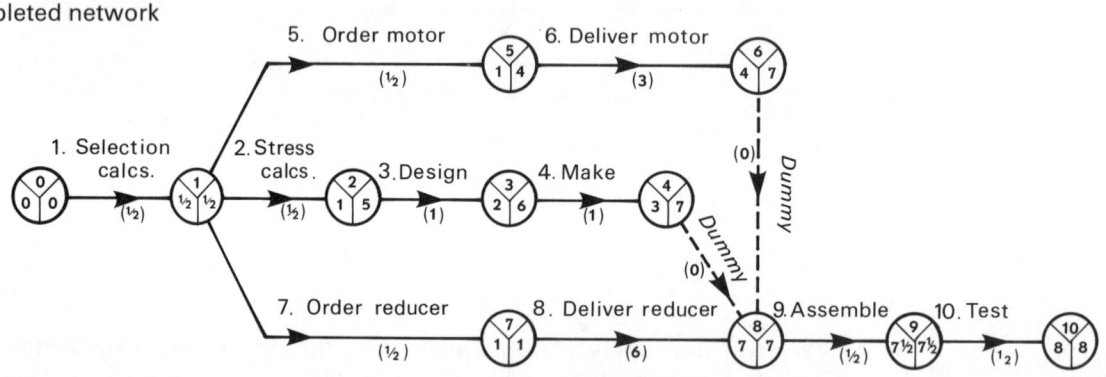

Activities (time in weeks)

1. Design calculations (select motor, coupling and speed reducer) (½)

2. Stress calculations (½)

3. Design bedplate and guard (1)

4. Make bedplate and guard (1)

5. Order motor (½)

6. Deliver motor (3)

7. Order speed reducer (½)

8. Deliver speed reducer (6)

9. Assemble (½)

10. Test (½)

Assembly (9) cannot commence until: the fabrications are manufactured (4); the motor is delivered (6); and the reducer is delivered (8). Dummies thus converge from circles (6) and (4) onto circle (8) on the critical path.

Latest completions for low-priority tasks are the last things to be entered on the network, e.g.

Latest completion for activity $(6) = 7 - 0 = 7$ weeks

Latest completion for activity $(5) = 7 - 3 = 4$ weeks

The Complete Planning Exercise

One popular planning system is known as PERT (Programme Evaluation and Review Technique) and was developed in the late 1950s on the complex Polaris fleet ballistic missile programme for the U.S. Navy.

The process is as follows:

1) List the activities, and estimate times.
2) Construct a network.
3) Prepare an *activity sequence chart.*

Fig. 2.6 gives an illustration of the PERT technique.

As an example, consider that a design manager wants to schedule a project for a general-purpose gear pump. Fig. 2.7 shows the activities as a schedule and progress record. The sequence of activities need not be considered at this stage. Fig. 2.8 shows the completed network.

The activity sequence chart is now drawn up. This represents the complete project and contains the names of individual staff to whom each activity is assigned. Manpower can thus be planned. The activities are placed in the order obtained from the network. Fig. 2.9 shows the completed activity sequence chart. The design manager has, in this case, allocated three members of the staff to the project in order to keep the total time to what is considered a minimum.

During the process of planning, the design manager can move staff around between various projects in order to meet the critical path times.

Fig 2.6 The PERT technique

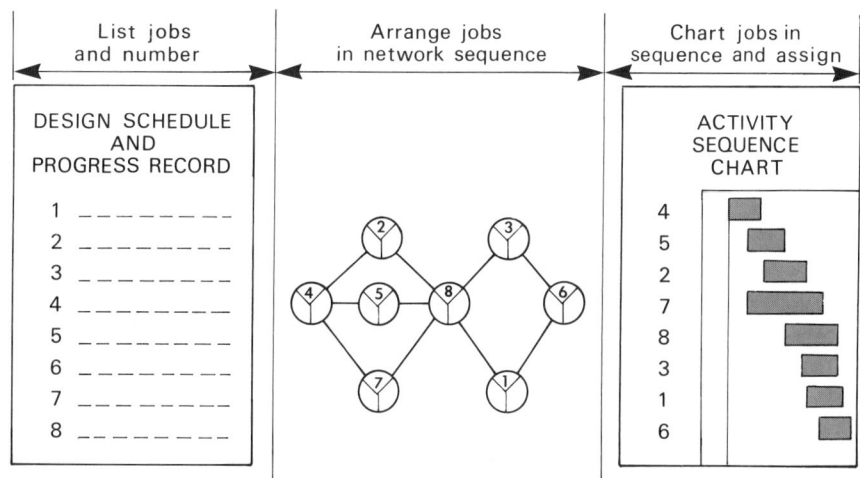

Fig 2.7 Schedule and
progress record

DESIGN SCHEDULE AND PROGRESS RECORD

PROJECT : GEAR PUMP FOR GENERAL-PURPOSE USE

JOB	ASSIGNMENT	ESTIMATED TIME (HRS)	ACTUAL TIME (HRS)
0	Begin project	0	
1	Calculate stresses	12	
2	Draw housings	30	
3	Draw gears	5	
4	Draw shafts	6	
5	Scheme pump	12	
6	Check details	10	
7	Draw up parts list	5	
8	Issue drawings	2	
9	Select material	2	
10	Issue parts list	1	
11	Design release	0	
	Total	85	

Fig 2.8 Completed
network

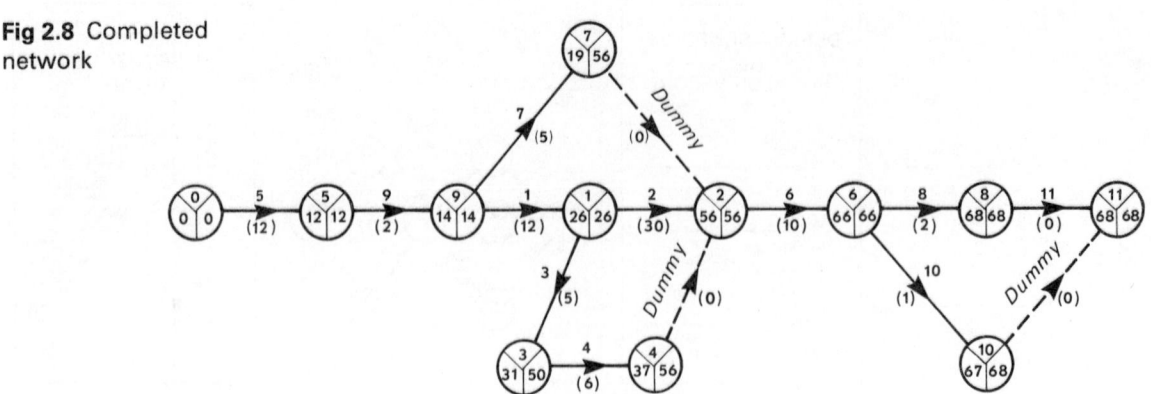

Fig 2.9 Activity sequence chart

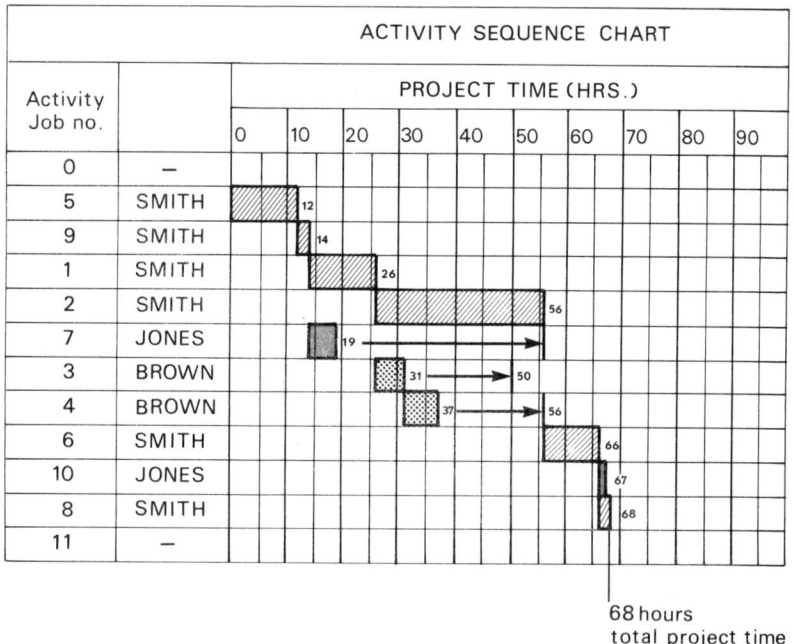

68 hours
total project time

Sometimes during the course of a project it is necessary to increase the number of people working on it if the design manager sees the project falling behind schedule. Conversely, staff could be moved off a project which is ahead of schedule to one which is behind, or deemed to be of greater importance. It is a mistake to assume that a planning procedure must be rigidly adhered to. The best-laid plans can come adrift due to unforeseen circumstances such as sickness, design problems, change of customer requirements, etc. The project must be continually studied to ensure that it meets the schedule times and that none of these problems have been allowed to cause a deterioration.

**2.3
Design
Organisation
Structures**

The organisation of the drawing and design office will depend on the size of the company. It is not necessarily dependent on the overall structure of the company itself. A regular structure is essential to maintain continuity of work and to ensure the appropriate discipline amongst the workforce. Also, it can be seen that promotion is possible through the orderly structure.

Fig. 2.10 shows a structure for an office dominated by one person, the boss. A regimented office structure has been established which will be very disciplined and able to perform specific well-defined tasks. This is the normal type of basic organisation structure. Modification to it changes the effectiveness of the organisation.

Fig 2.10

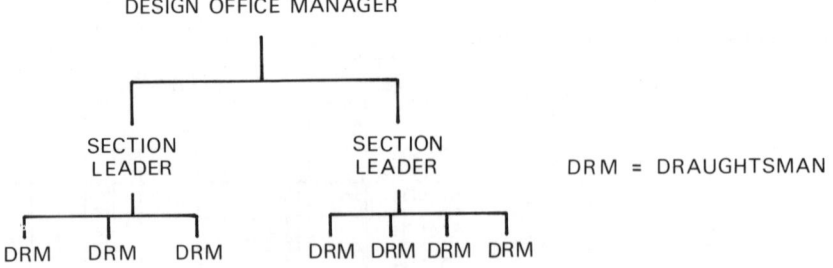

Fig 2.11

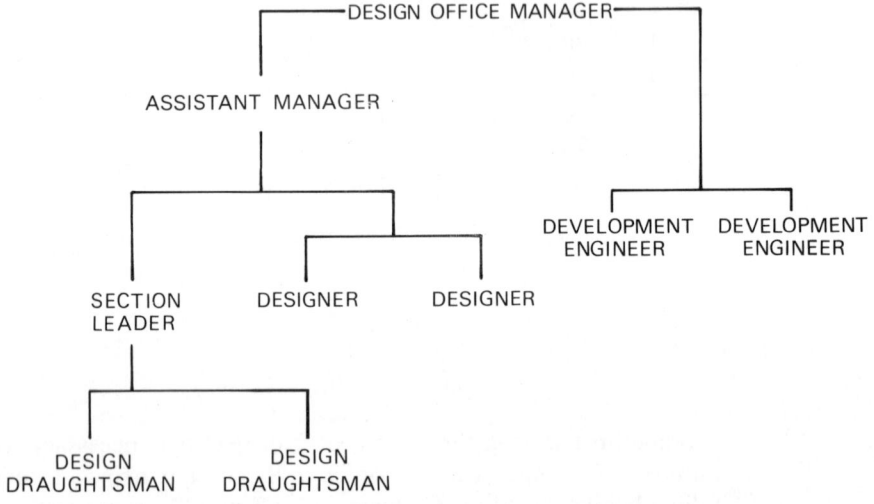

Fig. 2.11 shows an office structure divided between the design and development functions of the company. This type of structure allows the manager to concentrate on the activities in the development of the products produced through the drawing office. The drawing office is basically run by the assistant manager who would only report to the Design Manager on aspects of policy and would deal with day-to-day running of the office completely. This type of structure would be more suitable for the smaller company where the design office is responsible for both design and development functions.

Larger Structures

Fig. 2.12 shows the type of structure which might exist in a large organisation. The Engineering Director is the figurehead of the structure and is only consulted, normally, on affairs of policy and major problems. The normal day-to-day running of the department is carried out by the engineering manager who would report directly to the chief engineer. The chief engineer is responsible for liaison with the customer, formulating the work of the engineering manager, and, if highly technical, may put up design solutions for the engineering manager to progress through the department.

Fig 2.12

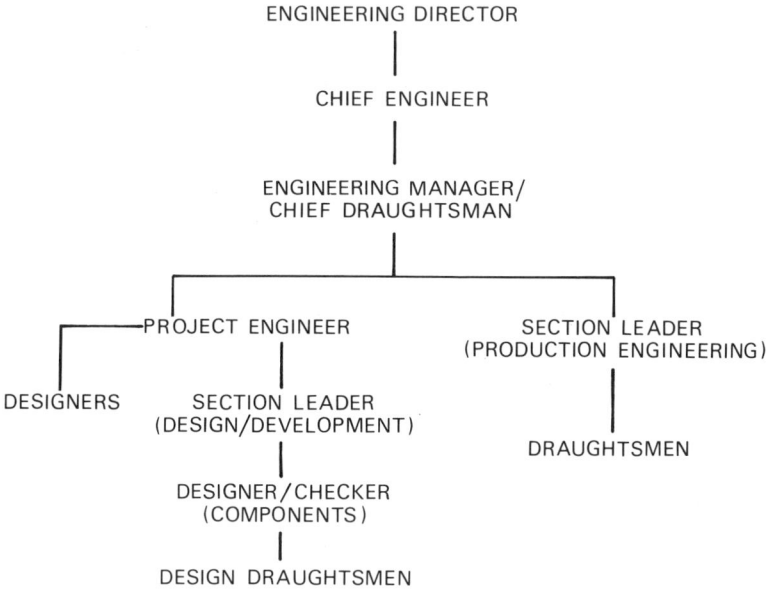

Fig 2.13

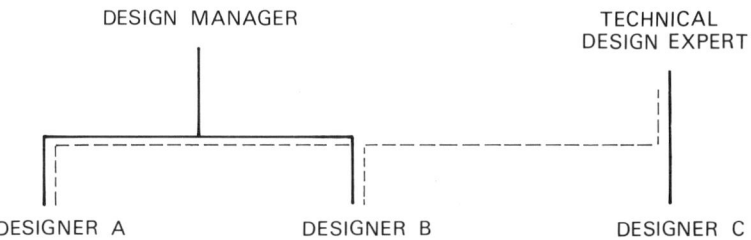

Matrix Structure

Fig. 2.13 shows an arrangement which occurs more and more nowadays to cater for the situation where a manager is required to organise the daily running of the department, and a technical expert is necessary for the high technological complexity of the work. In these situations a design manager may judge the adequacy of the technical aspects of the work and also be responsible for the training and other technical functions, whereas the technical design expert is responsible for the designers' work but not the personal problems that occur in the office.

This structure works well where there is a large amount of project work, for example in the case of a civil engineering department concerned with a process engineering project. Here the designers would need guidance from the technical design expert, but would go to the project manager for aspects of time scales, costing, salaries, etc. It is necessary to ensure that all the members of the office are well aware of their position within the office, that for instance problems of a personal nature may need expert attention from the project manager, otherwise the structure is unworkable.

Circular Structure   Fig. 2.14 shows an unsatisfactory structure which would only work for a short period of time. It could be used when a project is commenced in order to ensure that all departments work together in harmony. Once working, then each department reports to the director only on matters of project policy. As the project develops, then the individuals in each section become more aware of their individual responsibilities and need to depend on the director less and less.

2.4
Types of
Company

It was stated above that the design department structure is dependent on the size of the company. It is, however, also dependent on the type of product.

Large companies producing several products to suit more than one customer application demand a well-organised structure in the drawing office employing engineers, draughtsmen, technical clerks, etc., who are specialists in certain disciplines. Conversely a small company whose profitability depends on one product will require a small drawing office. In this case the designer would be responsible for design detailing, specification, feasibility studies, costing, organising the purchase of individual bought-out finished items, clerical work, and customer liaison on the product design and application. Such work is far more demanding but personally rewarding. It is also apparent that such a designer's knowledge of materials, processes and organisation must be superior to that of a designer working in a large organisation structure.

An example of the large-structure organisation in the drawing office is found in the manufacture of motor vehicles. A smaller drawing office structure may be found in the manufacture of, say, garden shears or small agricultural implements. Examples of the two types of drawing office structure are given below, in terms of the type of product.

Large-organisation structure	*Small-organisation structure*
Tractors.	Circlips.
Fuel Pumps (car/truck).	Hose Clips.
Aircraft.	Door Locks.
Computers.	Electric Kettles.
Engines.	Small domestic products
Bearings.	(e.g. curtain rails).

2.5
Design Groups

It can be advantageous to split the design department into individual product groups, especially where the company is involved in developing several products at the same time. In cases like this it is normal to have a team leader who has the overall responsibility of organising these product groups, getting the best ideas from them, and then formulating a work programme.

Each team member should be expected to individually prepare for group discussion a series of preliminary ideas on the design of the piece of equipment for which they are responsible. A complicated design problem cannot be solved by a single engineering designer or specialist since many disciplines are represented.

Fig 2.14

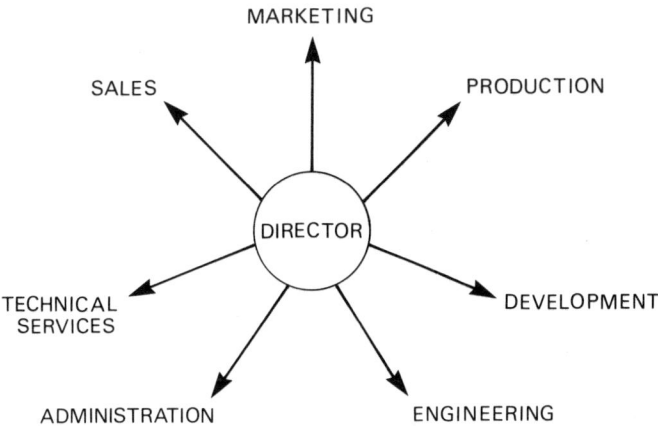

Fig 2.15

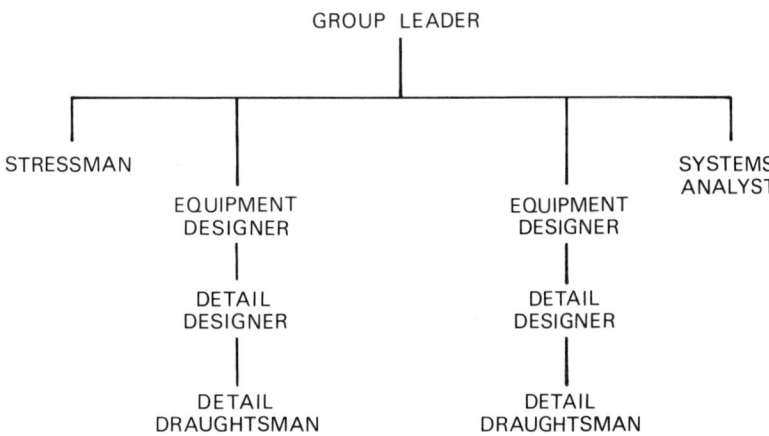

A typical team may consist of the personnel shown in Fig. 2.15. The group consists of a leader and two equipment designers backed up with a stressman and systems analyst. Together this group of specialists can be a formidable team and able to problem solve very effectively.

Individual Functions

The team members as shown in Fig. 2.15 each have a predetermined function so that they can cach operate effectively. Their functions are briefly stated below.

STRESSMAN Because all designs have some form of mechanical limitation in respect of load-carrying capacity, it is necessary to ensure that the design does not fail due to adverse stress.

The stressman is an expert in determining the design limitations and calculating the stress imposed due to loading conditions. However, the designer is responsible for guiding the stressman along the correct lines and outlining the possible load conditions that could exist.

Typical items the stressman must consider are

1) *Design for stiffness*: for example, to make sure that high-speed shafts are able to carry the torque.
2) *Design for strength*: for example, to make sure that bolts holding two parts together are strong enough to do so.
3) *Design for abuse*: for example, to make sure that machine components are strong enough to withstand shock loads due to incorrect use.

EQUIPMENT DESIGNER   Responsible for the overall design of a piece of equipment; will prepare the original design layout; can undertake the initial calculations or pass instructions to the stressman.

SYSTEMS ANALYST   Responsible for making sure that a system will work. For example, the hydraulic or pneumatic circuit of a machine tool must be feasible and the systems analyst would prepare calculations, possibly using a computer, to check that this is the case. This person would normally be a qualified development engineer and carry out these investigations separate from normal duties.

DETAIL DESIGNER   Responsible for the design of individual component parts, taking account of form, size and method of production.

DETAIL DRAUGHTSMAN   Responsible for detailing small parts, and parts which have been evaluated by the detail designer.

The engineering designer needs to be qualified in many disciplines, with an excellent knowledge of mathematics, physics, geometry, manufacturing technology and engineering design.

A detail designer, however, would need only a basic professional training in technical drawing, design and mathematics.

The detail draughtsmen are normally enthusiastic juniors who are going through a course of training, the in-house portion of which is provided by the detail designer, concurrent with a college education covering the required groundwork of mathematics, workshop technology and technical drawing. Progress could eventually lead to designer status.

The profession of Engineering Designer is one which requires many years of training and experience. In addition to a thorough understanding of design techniques, the engineering designer will also require considerable knowledge in a wide range of specialisms. These may include: science and mathematics; production engineering; ergonomics (study of humans and their relationship with machines); aesthetics (concerning the appearance of designs); legal aspects (e.g. safety requirements). The profession also demands many personal qualities. Inventiveness and creativity are essential. Other required qualities include: ability to think logically; ability to communicate ideas; a good memory; a good sense of responsibility. In fact the engineering designer never stops acquiring diverse knowledge and skills. Of paramount importance is a willingness to keep abreast of modern technology by reading technical journals and manufacturers' literature and by attending seminars.

Fig 2.16

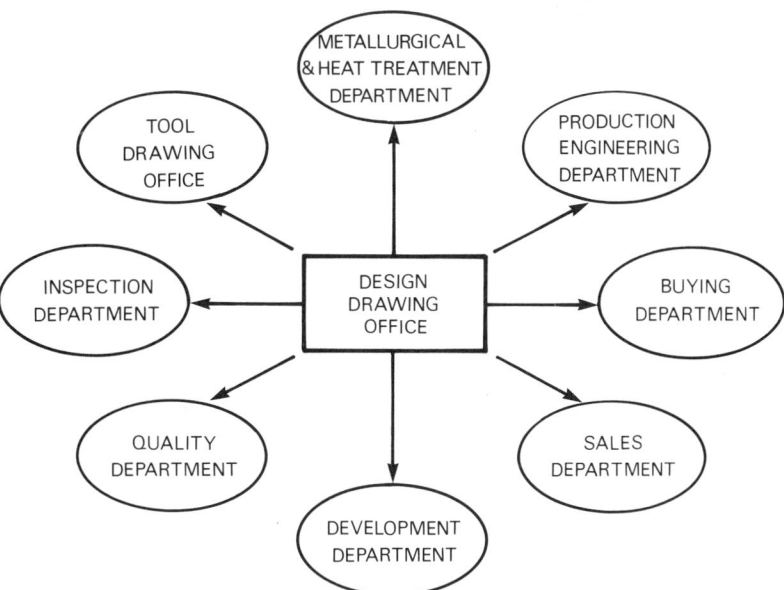

Inter-relationships between Departments

Fig. 2.16 depicts the design drawing office as the hub of the company's business. The interchange of ideas is essential. Only by bringing together the responsible engineers from each of the departments can progress be made towards achieving the best product in the way of function, cost, material, customer appeal and form. It is particularly important to maintain good liaison between design and production departments so that the intentions of the designer are known to the production engineer. Early discussions here will result in benefits such as cost savings by choosing the appropriate component shape, optimum production technique, and minimum required machining.

**2.6
Stages of
Evolution of
the Product**

There are several very important stages in the evolution of a product. These can be listed as

    1) Initiation
        a) Conception
        b) Feasibility
    2) Design and Development
    3) Pre-production
    4) Production.

Before a problem can be solved it must first be identified. For example, suppose in the manufacture of a certain plastics material it is necessary to oscillate the processing machinery. This oscillation must be produced by a mechanism which is drawn by a rotating shaft (see Fig. 2.17). The shafts A and B must be at right angles to each other, and in the same plane. The problem has therefore been identified; it remains to analyse it, put up various solutions as stated in section 1.1, and then carry out a synthesis of the problem.

Fig 2.17

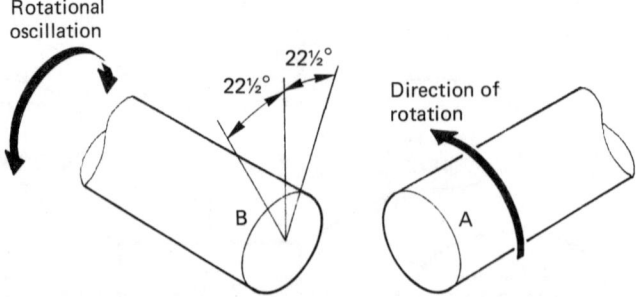

Feasibility

It is essential for the design to be a feasible proposition; in other words the final design solution should be a practical proposition. Feasibility studies are defined as:

a) A cycle of design steps which validates the need, defines the problem, and appraises possible solutions.

b) It also examines whether or not a design solution will function to the given specification at an economic cost.

Design and Development

These two are closely related activities. Having decided on which proposal to proceed with, the designer prepares layouts and supervises draughtsmen in the preparation of detail drawings. The development engineer is responsible for ensuring that the product's performance meets with the original specification, and functions correctly.

Pre-production

Having progressed the product through the development stage, certain features may have to be altered to fulfil correct functioning and to ease production of the parts.

This is normally the period when the designer discusses the drawings with the production engineers. From these discussions, the detail drawings are "productionised". For example

a) Tolerances are adjusted to suit machining processes.

b) Component form (shape) is modified to suit manufacturing processes.

c) Radii and sharp edges, etc., are modified to suit tooling and processes.

A batch of assemblies is then produced which conform to these modified drawings and which are used to anticipate any likely problems when full production commences.

Production

Having established that it is feasible to manufacture the component parts in large quantities, then the production department prepares final machining layouts and the tooling departments organise the necessary tools and fixtures suitable for producing the parts. Machine tools are allocated to produce the individual parts and, if it has been found necessary, new ones are ordered.

Problems

2.1 What are the methods used for project planning? Explain why they are used.

2.2 Produce a Gantt chart for the design of a worm gearbox consisting of the following items

1) Housing   5) Worm Wheel
2) Cover Plate   6) Oilseals
3) Bearings   7) Bolts
4) Worm Shaft   8) Washers

Include details of how you propose to allocate two people to produce these drawings, parts list and drawing issue.

2.3 It is required to extend the work in producing the worm gearbox in problem **2.2**. The following assignments are required to be completed up to the issue of drawings:

Calculate stresses.
Scheme gearbox.
Draw component parts.
Prepare parts list.
Issue drawings and parts list.

Include any other activity you think is required. Estimate times to complete these assignments and prepare

a) a design schedule and progress record
b) a network
c) an activity sequence chart.

2.4 *a*) Prepare an organisation structure of your own place of work

b) Prepare an organisation structure for your company's drawing office.

State the type of structure you think it is, and say whether you think that by minor reorganisation it could be made more efficient.

2.5 Prepare a structural organisation chart for your design group and state each team member's function within it.

2.6 What qualities and knowledge does a designer in your company require to ensure the proper design of the product produced there.

2.7 Explain in your own words how you would initiate a design, develop the product, and pass through to pre-production stage. Use your own company as a basis for your answer.

3 Design for Economic Manufacture

3.1 Costs and Value

It is important that the designer develops skills in estimating costs of designs in order to secure the most economic features. This entails being aware of the breakdown of product costs and Fig. 3.1 shows the various cost stages that exist in an organisation. These have to be broken down into individual units to calculate the overall cost of the product.

For most purposes of calculating costs however, it would be extremely difficult, and time consuming, to put together these individual costs, but they can be represented by three basic unit costs which are:

a) Material costs.
b) Labour costs.
c) Overheads.

a) plus *b*) is the unit cost of manufacture, and *c*) is the cost of preparation of the component.

The term "cost of producing the product" is ambiguous because it depends upon the quantity produced.

Fixed Costs and Variable Costs

Some costs vary with output quantity, whereas others are said to be fixed because, at least over a fairly considerable range of output, they do not vary with output. For example, once a company has built its factory and installed the necessary machinery, these costs remain the same whether the company is working at full or part capacity. The amount of rent and rates is near enough constant and the number of staff employed may not vary significantly with output. Fixed costs also embrace property tax, interest, insurance and research costs. Therefore it is important that the project is carefully costed prior to manufacture and that the designer bears some responsibility for the outcome of these costs.

Material costs come under the heading of variable costs, as do labour, wages and heating, lighting and machinery power bills. These can vary considerably from the time a project commences until the project is completed and should be checked at regular intervals to ensure that the budgeted values are not exceeded.

The Selling Price

The major factors affecting a company's pricing policy and which are crucial in maintaining a healthy business are:

a) Value
b) Supply
c) Demand.

The value of a product is closely related to the design itself, especially if innovative in concept. The customer looks for value for money when goods are ordered from a manufacturer, and it is up to the manufacturer to ensure

Fig 3.1 Breakdown of product costs

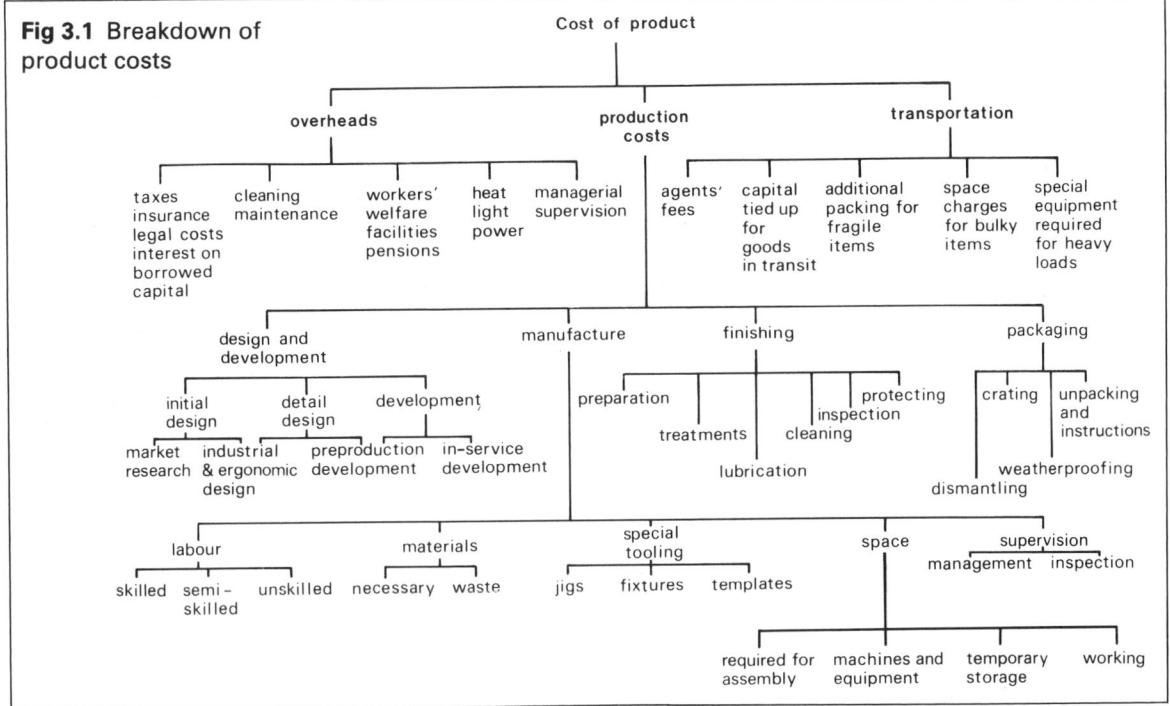

Fig. 3.1 is reproduced courtesy Dr. G. Pitts, Lecturer in Mechanical Engineering Design, University of Southampton

that the goods do meet certain standards. These standards can be met by installing equipment for measuring the *quality* of the parts. Other items which can affect the value of a product are:

a) Reliability
b) Availability
c) Delivery period
d) After-sales service.

All goods manufactured and sold are governed by the market trend. If a product is readily available from several companies (i.e. supply is high), the selling price will generally be low in order to increase competitiveness. Low supply, however, may induce high demand which tends to generate high selling prices. Thus it becomes necessary to equate demand with the supply of a commodity.

3.2 Design and the Marketplace

Design distinguishes a product from competitive products. To make an impact on the marketplace, the designer needs to present novel, innovative and well-balanced designs in terms of aesthetic appeal and function.

The customers' individual requirements will vary considerably and manufacturers have to decide upon the market they intend to aim for. It is well known that some manufacturers are very good at going for the lower end of the market. For example, some car manufacturers are able to fit into this end by reducing the complexity of cars and giving the basic requirements of decor, function and design appeal without losing the benefits of good safety and mechanical design.

On the other hand, Rolls Royce can only produce cars of elaborate distinction because that is their specialism. Therefore they aim their products at the top end of the market knowing that, although a smaller demand exists, their pricing policy will ensure healthy business activity and survival.

Design is an investment, not an overhead which can be cast aside as a second-rate activity. This is borne out by the increasing use of computer-aided design and draughting techniques in industry today. It has been realised that without good design facilities it will not be possible in the future to be competitive, and such companies will be priced out of business.

3.3
Costing Designs,
Surface Finish
and Production

In costing designs, it is important to remember that the cost of a component will depend as much on the method by which it is manufactured as on the types of material chosen for it.

The type of surface finish produced by various machining processes is shown in Fig. 3.2a. This shows that several processes are suitable for producing 0.4 μm finish. Therefore it pays to examine this type of chart and choose the most economic machining process.

Fig. 3.2b shows the approximate relationships which exist between relative cost and surface finish. To choose a 0.4 μm finish instead of a 1.6 μm finish would result in the part being approximately 1.5 times more expensive to produce. It is therefore important to select the correct machining process and also the correct surface finish.

The actual production cost of a component is made up of
a) Unit cost of manufacture
b) Set-up cost.

The unit cost of manufacture accounts for the method of manufacture and costs incurred thereby. This may include costs of fettling, chamfering, de-burring, or any similar operation. The set-up cost is the cost incurred by the method of placing parts in a machine, die, mould or any similar method. The total cost per component may be summarised in the formula:

$$T = \frac{C}{Q} + U$$

where T = total component cost
C = cost of preparation
Q = batch quantity
U = unit cost of manufacture.

It should be remembered that the costs shown in Fig. 3.1 are all included somewhere in the total, but for ease of calculation they are amortised into the final values of unit cost and set-up cost.

3.4
Cost/Income
Comparisons

The "profitability" of a design is particularly affected by the number of articles produced and may be investigated by comparing production costs with the expected income from a realistic selling price. If the investigation reveals a poor chance of profitability, the design may need to be modified, or a different production technique considered. To assist in this process a

Fig 3.2*a* Surface finish chart

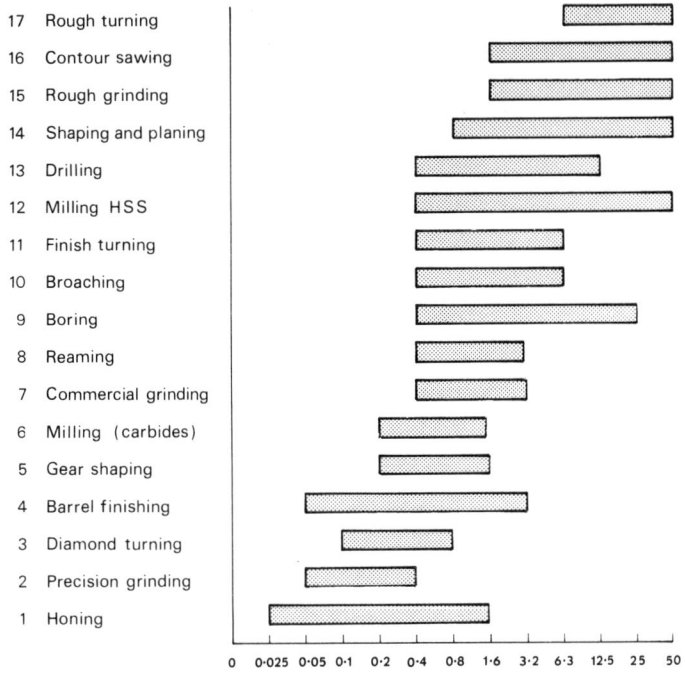

17 Rough turning
16 Contour sawing
15 Rough grinding
14 Shaping and planing
13 Drilling
12 Milling HSS
11 Finish turning
10 Broaching
9 Boring
8 Reaming
7 Commercial grinding
6 Milling (carbides)
5 Gear shaping
4 Barrel finishing
3 Diamond turning
2 Precision grinding
1 Honing

0 0·025 0·05 0·1 0·2 0·4 0·8 1·6 3·2 6·3 12·5 25 50

Surface finish (μm)

Fig 3.2*b* Surface finish and relative cost

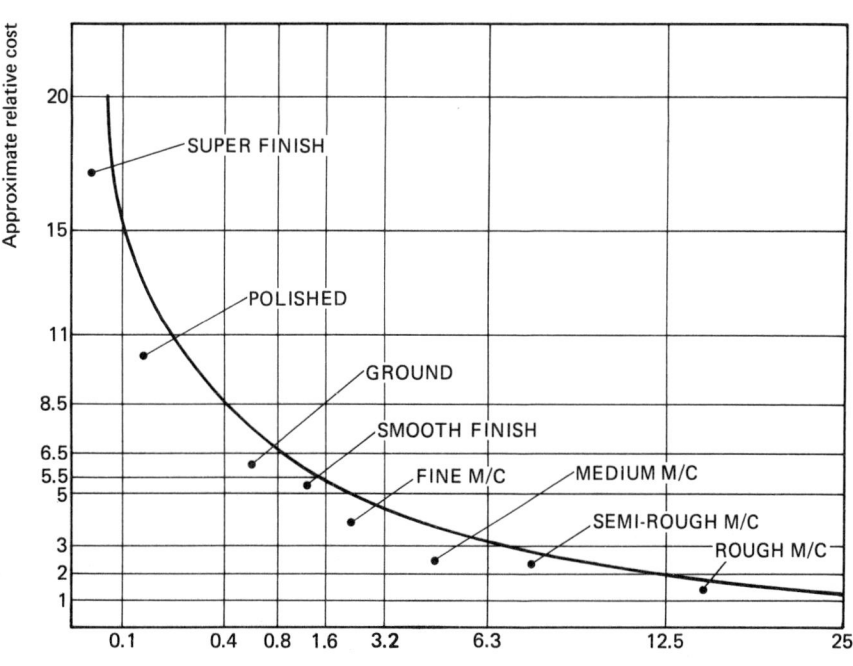

Surface finish (micrometres μm)

Break-even Chart is often used. This plots production costs and expected income, against the quantity to be produced.

Fig. 3.3 shows the basic break-even chart. AB represents the fixed costs which are not affected by the output quantity. In theory, these remain constant, although adjustments may be necessary to account for inflation.

Fig 3.3 Basic break-even chart

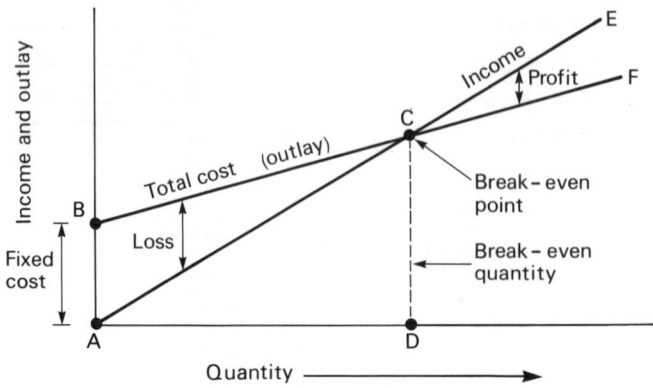

The line BF represents the variation of total cost expected against output quantity. The line AE represents the variation of expected income against output quantity. At point C, where these two lines cross, income exactly equals the cost of production, and this is called the Break-even Point. Point D reveals the quantity which must be produced in order to break even. Any parts produced after this point will result in a profit.

Production Cost Comparisons

The break-even chart may also be effectively used to compare the cost of two or more production methods for a given design.

Consider the example of producing the bush shown in Fig. 3.4. The part could be made on

a) A centre lathe
b) A capstan lathe
c) An automatic lathe.

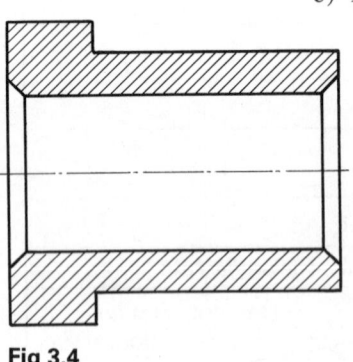

Fig 3.4

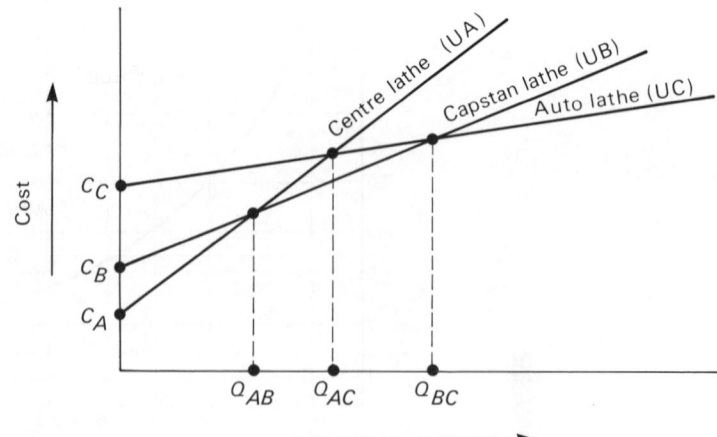

Fig 3.5

Fig. 3.5 shows the break-even graph which is used to determine which production method will be the most economical. As can be seen, there are three break-even points and the method of finding which one gives the most economical production method is to use the following formulae.

A For the Centre Lathe

$$\text{Total component cost} = \frac{\text{Cost of preparation}}{\text{Batch quantity}} + \frac{\text{Unit cost of}}{\text{manufacture}}$$

$$T_A = \frac{C_A}{Q} + U_A$$

B For the Capstan Lathe   **C** For the Auto Lathe

$$T_B = \frac{C_B}{Q} + U_B \qquad\qquad\qquad T_C = \frac{C_C}{Q} + U_C$$

Referring to Fig. 3.5, a cost break-even point is reached when the lines for producing by centre lathe and by capstan lathe cross at a quantity designated Q_{AB}. Below this batch quantity, it is cheaper to produce on the centre lathe.

This break-even point occurs when

$$\frac{C_B}{Q_{AB}} + U_B = \frac{C_A}{Q_{AB}} + U_A$$

$$Q_{AB} = \frac{C_B - C_A}{U_A - U_B}$$

In other words the break-even point is calculated by dividing the difference in set-up costs by the difference in unit costs.

Similarly

$$Q_{BC} = \frac{C_C - C_B}{U_B - U_C} \quad \text{and} \quad Q_{AC} = \frac{C_C - C_A}{U_A - U_C}$$

This process can also be used for selecting between various methods of fabricating, casting, moulding, etc., and between these and alternative methods.

The bush example has been used to identify the way in which three options can be compared. We shall now look at a working example.

3.5
Cost Comparison:
a Worked Example

Consider the case of producing the handle on a plastic bucket. Essentially it is a piece of plastic tubing which can be made in a variety of ways. However, only two ways will be compared:

a) Parting-off tubing on an automatic lathe.
b) Moulding the handle complete.

The demand for the buckets to which this handle is to be attached is limited to about three months of the year. Assume that the share of the market is 15 000.

Method A

Set-up cost $C_A = 1$ hour @ 700 pence/hour

$$= 700 \text{ pence}$$

Material cost of extruded tube $= 0.07$ pence/handle
The operation of radiusing the handle ends and parting-off the tube takes 0.1 minute at 700 pence/hour, i.e.

$$0.1 \times 700 \times \tfrac{1}{60} = 1.167 \text{ pence/handle}$$

Total unit cost $U_A = 1.167 + 0.07$

$$= 1.237 \text{ pence}$$

Total component cost for batch of 15 000 handles is

$$T_A = \frac{C_A}{Q} + U_A = \frac{700}{15\,000} + 1.237 = 1.284 \text{ pence}$$

Method B

Set-up cost $C_B = 5$ hours @ 700 pence/hour

$$= 3500 \text{ pence}$$

Material cost of plastic granules $= 0.014$ pence/handle
The moulding operation produces 300 shots per hour from a four-cavity mould, i.e. 1200 handles/hour.
Moulding labour charge $= 700$ pence/hour
Therefore, each handle costs

$$\frac{700}{1200} = 0.583 \text{ pence}$$

Total unit cost $U_B = 0.014 + 0.583$

$$= 0.597 \text{ pence}$$

Total component cost for batch of 15 000 handles is

$$T_B = \frac{C_B}{Q} + U_B = \frac{3500}{15\,000} + 0.597 = 0.830 \text{ pence}$$

Then, the quantity of handles at which it becomes more economical to mould complete instead of parting off tubing is

$$Q_{AB} = \frac{C_B - C_A}{U_A - U_B}$$

$$= \frac{3500 - 700}{1.237 - 0.597} = 4375 \text{ handles}$$

Fig. 3.6 shows how to find this quantity by a graphical method.

Fig 3.6

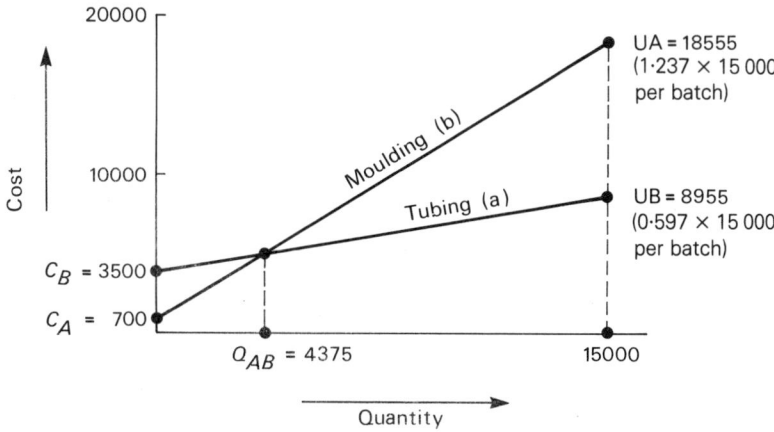

Margin of Safety

It is necessary for budgeted income to exceed break-even income by as much as possible; this ensures a healthy business. The margin-of-safety figure can be calculated as follows:

$$\text{Margin of Safety} = \frac{\text{Budgeted income} - \text{Break-even income}}{\text{Budgeted income}} \times 100\%$$

For example, if

Budgeted income = £20 000
Break-even income = £16 000

then

$$\text{Margin of safety} = \frac{20\,000 - 16\,000}{20\,000} \times 100 = 20\%$$

The product would have a margin of 20% over which the company could afford to be in error due to unknown events.

3.6
Economics of Design

The following considerations highlight some of the methods by which thoughtful design may reduce the overall product cost.

1 *Economic Choice of Production Technique* As already discussed, this is governed largely by the quantity produced and the available plant of the company.

2 *Economic Choice of Material* Table 3.1 shows how the costs of various materials differ. The table also shows typical densities for each material. This gives a good indication of the weight saving that can be achieved by the choice of alternative materials. Of course, the cost of a material must always be balanced against its mechanical properties. For example, a stronger material may result in smaller component sections and thus become more economically viable than a cheaper, weaker material.

The type of data shown in Table 3.1 will obviously alter with changes in availability and processing techniques, and should thus be regularly reviewed.

3 *Economic Choice of Design Form* Section 3.3 raised the subject of surface finish and how it affects the cost of producing an item. Other aspects that must be considered include the form or shape of the final design. This must be kept as simple and light as possible, and yet still have adequate strength to sustain working loads. Chapter 5 deals in some detail with the aspect of strength/cost optimisation of design form with emphasis on strength/mass ratios and stiffness/mass ratios.

4 *Avoiding Excessive Machining and Close Tolerancing* Excessive use of close tolerances and geometric tolerances adds cost to the design. The need to set up special inspection and quality control techniques should be minimised, as this not only increases direct costs, but it can also lead to a way of creating scrap parts through improper handling.

The designer can minimise the amount of machining by raising the surfaces that need to be machined as shown in Fig. 3.7.

Fig 3.7 Examples of minimising the amount of machining by raising surfaces

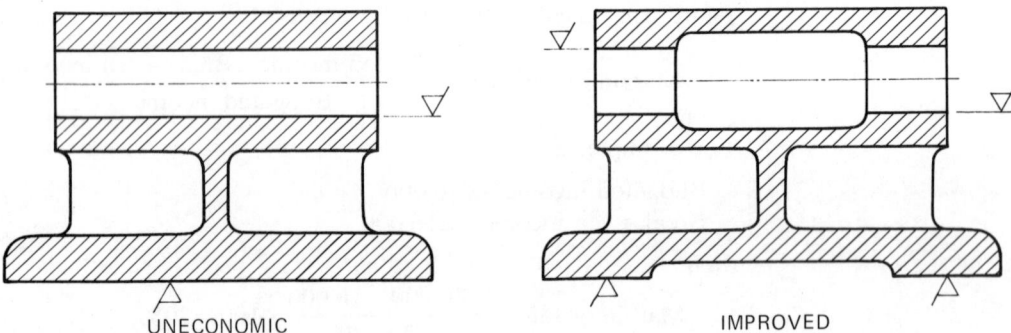

UNECONOMIC   IMPROVED

5 *Avoiding Material Wastage* Most production techniques involve some wastage of material, especially where machining is involved. However, a simple design modification can sometimes greatly reduce this effect, as shown in Fig. 3.8

Fig 3.8 Example of minimising material wastage

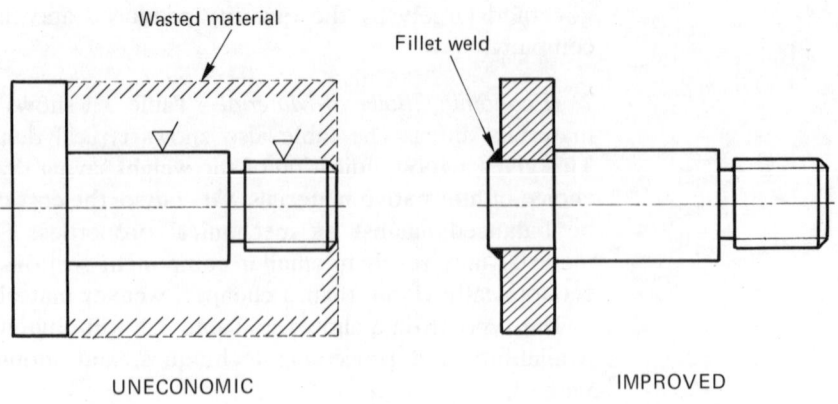

Wasted material

Fillet weld

UNECONOMIC   IMPROVED

Table 3.1 Comparison of material costs

Material	Density (g/cm^3)	% cost/kg 220 M07 (EN 1A)	% cost/cm^3 220 M07 (EN 1A)
PLASTICS			
Vinyl	1.30	147	25
Polystyrene	0.92	156	19
Fibreglass	1.61	226	47
A.B.S.	0.83	339	36
Nylon	1.14	739	109
IRONS			
Grey-iron casting	7.10	139	128
Ductile-iron casting	7.10	278	256
Malleable-iron casting	7.30	330	313
Steel casting	7.91	348	356
DIE CASTINGS			
Zinc die casting	6.66	365	316
Aluminium die casting	2.66	478	166
Brass sand casting	8.88	913	1046
Manganese bronze casting	8.21	1391	1481
STEEL			
Sheet steel 220 M07	7.85	59	59
Bar steel 220 M07 (EN 1A)	7.85	100	100
Steel forging	7.85	304	309

6 *Consideration for Size and Product* Further cost savings can be made if due regard is given to the size of the product. For example, in Fig. 3.9, by putting holes through the bracket, the mass of the part is reduced and material costs are reduced. If the part is small, however, Fig. 3.10 shows that no real advantages accrue by putting cored holes in the bracket since the section around the holes would have to be strengthened, and also more problems would exist in handling the part.

7 *Efficient Use of Standard Components and Bought-out Items* Where knobs, handles, bearings, etc. are to be incorporated in the design, then it is advisable to consult catalogues of manufacturers which specialise in these parts. It may be far cheaper to purchase such items than to attempt unfamiliar production processes—especially if the required quantities are low.

8 *Designing to Aid Packaging* Packaging materials, haulage space, and freight charges are important cost considerations which may be affected by the shape of the finished product. Rounded shapes tend to take up more packaging space than rectangular or triangular, as shown in Fig. 3.11. Severe

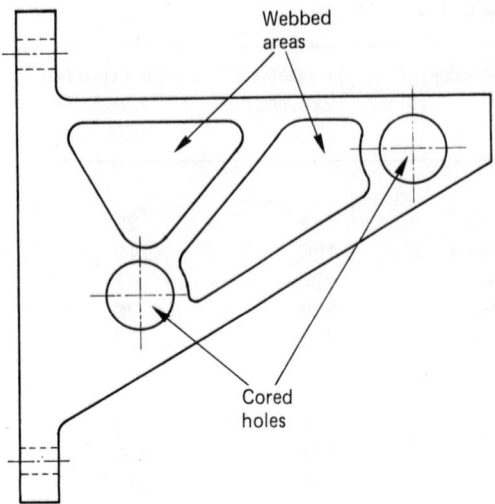

Webbed areas

Cored holes

Fig 3.9 Example of cost saving by minimising the amount of material required and the mass of the part

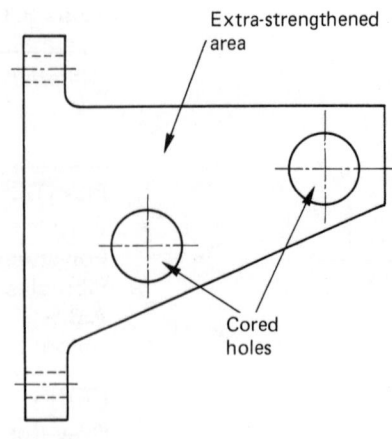

Extra-strengthened area

Cored holes

Fig 3.10 False cost saving

Fig 3.11 Designing for packaging

EQUAL AREAS

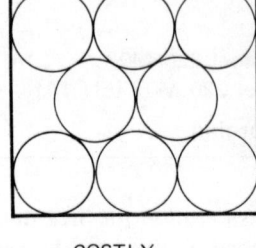

COSTLY PACKAGING

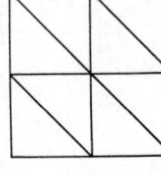

APPROX. 25% SPACE SAVING

Fig 3.12 Designing for packaging and ease of assembly

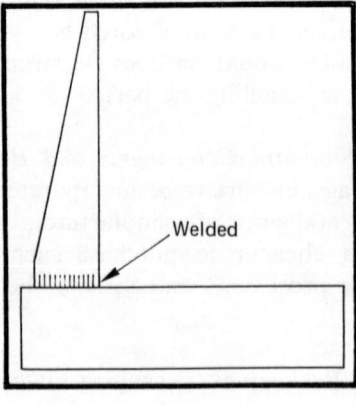

Welded

COSTLY PACKAGING

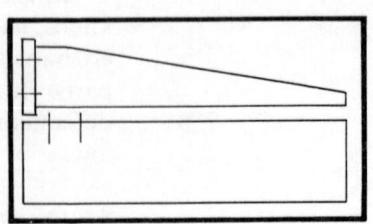

IMPROVED BOLTED COMPONENTS, DISMANTLED

protrusions are also space-consuming, and where these are unavoidable, the product should be designed for easily dismantling and re-assembly as shown in Fig. 3.12. A good example of this is also discussed in the design case study of Chapter 10.

9 *Using Computer-Aided Techniques* The advent of the microcomputer has now provided the designer with a tool for speedy and reliable analysis of the foregoing considerations. An excellent example of this is shown in Fig. 3.13. Other aspects of computer-aided design (CAD) are fully discussed in Chapter 9.

Fig 3.13 Redesign of zinc casting by CAD

Shown in Fig. 3.13 are two versions of a zinc die casting for a pair of small yokes. On the left is the "before" casting which used an excessive amount of molten material to fill its runner channels. On the right is the "after" version, redesigned by Lesney Industries with the help of a microcomputer program produced by the Zinc Development Association but adapted by Lesney for their own needs. By dramatically narrowing the runner channels, the casting material has been reduced from 110 g to just 27 g, thus vastly reducing heat energy requirements, tripling the production rate, and cutting the overall production cost by a massive 75%.

**3.7
Standardisation**

Even when preparing an original design, there are always certain features which can be resolved by the use of standard components. Standardisation is obligatory to the designer, and the following list shows the various groups which exist:
1) Basic standards dealing with abstract concepts, nomenclature, number sequences, screw threads, and limits and fits.
2) Standards dealing with engineering drawing practice, e.g. size of lettering, thickness of lines, spacing, etc.
3) Material standards.
4) Standard components, e.g. bolts, nuts, pins, operating elements, etc.
5) Standards dealing with power transmission components, bearings and gear tooth systems.
6) Standard fittings, pipework, mounting flanges and other static objects.
7) Standards dealing with safety requirements.

There are, of course, distinct cost advantages in using standard parts, especially where large-batch quantities are concerned. These are

a) Lower original cost of design.
b) No special tooling costs.
c) No development costs.
d) No planning required.
e) No special drawings required.

By using standard components, the designer has a wider choice of material finishes, accuracy of manufacture, etc., because the component manufacturer bears the cost of design and development of the parts.

Type Standardisation

This term is sometimes applied to a product that can be sold to many different types of industry—for example, products such as circlips, bearings, oilseals. However, it can just as well apply to complete units such as motors, gearboxes, clutches, etc. Each application requires a certain size of unit to transmit the necessary power, and, because of this, varying sizes of housings, shafts, gears and bearings, etc., will be required. For example, the size of an electric motor depends on the shaft diameter and it is this which needs to be varied.

Fig. 3.14 shows a typical chart which tabulates series, and shaft diameters for a type of electric motor. The power ratings are based on these values. A designer of a machine would look down the chart to find the power which satisfies requirements and would then specify that series number in the specification.

Fig 3.14 Example of standardisation chart

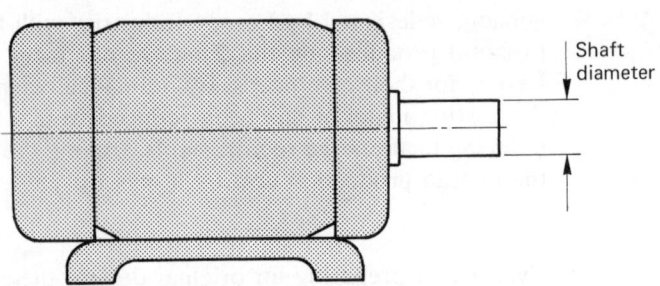

SERIES	SHAFT DIAMETER (MM)	POWER RATING
HA	10	
HB	12.5	
HC	16	
HD	20	
HE	25	
HF	32	
HG	40	

It will be noticed that the steps between each successive shaft diameter follow a pattern. This is a geometric progression and follows a form, 1, 1.25, 1.6, 2.0, 2.5, etc., where the common multiplier between each successive step is 1.25.

It has been found that a geometric progression satisfies human needs, and the designer should give due consideration to using it in situations such as this.

Geometric Progression formula
Progression $= a, ar, ar^2, ar^3, ar^4, \ldots$

$$\text{Common ratio } r = \sqrt[N-1]{\frac{x}{a}}$$

where $N =$ number of steps in series
$\quad\quad a =$ first number
$\quad\quad x =$ last number

Thus in the example of the electric motor:

$$N = 7 \quad\quad a = 10 \quad\quad x = 40$$

$$\text{Common ratio} = \sqrt[7-1]{\frac{40}{10}} = 1.2599 \simeq 1.25$$

Therefore, 10, 12.5, 16, 20, etc., have been selected for the shaft diameters.

Preferred Numbers

The series of standardised numbers known internationally as preferred numbers are very often used in arranging the sizes of standard components. This is done to achieve some measure of uniformity of size between a number of countries adopting this system.

The theoretical numbers, suitably rounded-off for practical convenience, are also known as Renard Numbers, after Colonel Renard who first proposed their use. The symbol R is used to identify them. The principal series are R5, R10, R20 and R40. These stand for numbers derived from geometric series having one of the common ratios which are

$$\text{R5: } \sqrt[5]{10} \quad\quad \text{R10: } \sqrt[10]{10} \quad\quad \text{R20: } \sqrt[20]{10} \quad\quad \text{R40: } \sqrt[40]{10}$$

Thus, the ratios are approximately equal to

1.58, 1.26, 1.12, 1.06 respectively

Table 3.2 shows the preferred numbers in the R5, R10, R20 and R40 series. Note that any series can be extended indefinitely upwards or downwards by multiplying or dividing repeatedly by 10.

Table 3.2 Preferred number series

Serial No.	Principal series				Serial No.	Principal series			
	R5	R10	R20	R40		R5	R10	R20	R40
0	1.00	1.00	1.00	1.00	20			3.15	3.15	3.15
1				1.06	21				3.35
2			1.12	1.12	22			3.55	3.55
3				1.18	23				3.75
4		1.25	1.25	1.25	24	4.00	4.00	4.00	4.00
5				1.32	25				4.25
6			1.40	1.40	26			4.50	4.50
7				1.50	27				4.75
8	1.60	1.60	1.60	1.60	28		5.00	5.00	5.00
9				1.70	29				5.30
10			1.80	1.80	30			5.60	5.60
11				1.90	31				6.00
12		2.00	2.00	2.00	32	6.30	6.30	6.30	6.30
13				2.12	33				6.70
14			2.24	2.24	34			7.10	7.10
15				2.36	35				7.50
16	2.50	2.50	2.50	2.50	36		8.00	8.00	8.00
17				2.65	37				8.50
18			2.80	2.80	38			9.00	9.00
19				3.00	39				9.50
					40	10.00	10.00	10.00	10.00

Problems

3.1 Explain in your own words how and why the designer needs to develop skills in estimating costs.

3.2 Why are the terms Value, Supply and Demand important to the health of the trading position of a company.

3.3 It is required to establish the break-even point of producing a link arm shown in Fig. 3.15. The designer has the choice of manufacturing the part from *a*) nylon moulding or *b*) aluminium die casting.

Fig 3.15

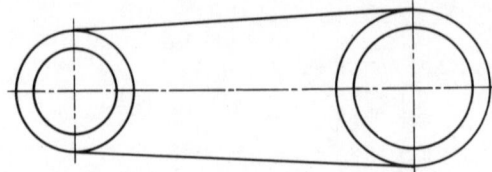

The intended production quantity is to be 18 000 units per year. Establish the material which would be the most economic choice for this quantity if

i) The set-up cost for producing the part as a nylon moulding is 5 hours at 700 pence per hour. The material cost is 12.2 pence.

ii) The set-up cost for producing the part as an aluminium die casting is 20 hours at 700 pence per hour. The material cost is 7.4 pence. The fettling cost is 1.2 pence.

[*Ans.* Break-even quantity = 2917
         Economic material for 18000 link arms is
         aluminium die casting]

3.4 If a product has a budgeted income of £350 000 and a break-even income of £200 000, what would be the margin of safety to allow for unknown trading factors.

[*Ans.* 42.9%]

3.5 Giving specific examples, describe typical design procedures by which the overall product cost can be kept to a minimum.

3.6 Part standardisation is essential in keeping a company's costs down. Explain how another company's products can be used to benefit by a designer.

Bibliography
C. V. Starkey, Cost break points, *IED Journal* (November 1969).

4 Dimensioning and Fits

This chapter outlines a systematic process of dimensioning which allows a designer to calculate unknown dimensions accurately and at the same time to ensure that the total allowable tolerance is shared between components in the correct proportion to allow for machining methods. The chapter assumes a basic knowledge of tolerancing and limits and fits as presented in *Engineering Design for Technicians*.

The importance of the correct tolerances on detail drawings has well established principles. Modern technological advancement, in addition to providing industry with improved manufacturing techniques, has led to increased demand for high productivity and manufacturing accuracy. It should be understood that

- *a*) Quality, reliability and value engineers have a need to receive drawings toleranced in a manner which simplifies manufacture.
- *b*) A product cost is directly related to the drawing tolerances.
- *c*) Excessive tolerancing means high factory costs by way of machine tools and inspection techniques.
- *d*) A systematic process of applying tolerances should be adopted.

4.1 Systematic Tolerancing

A system of vectors can be used to establish a suitable assembly condition. Fig. 4.1 shows how this applies. It is a typical dimensioning diagram which consists of a closed set of vectors. One vector represents the *dimension condition*, and the other vectors represent the dimensions *controlling* the dimension condition. The following symbol rules apply:

- *a*) The dimension condition vector is (+ve) and symbolised by $(+y_w)$.
- *b*) An example of a dimension vector in the positive (rightward) direction is $(+x_a)$ and in the negative (leftward) direction by $(-x_b)$
- *c*) The fundamental property of a dimension vector is that it goes between one surface and another *on the same part*.
- *d*) The surfaces at which the dimension vector begins and ends are surfaces that mate with adjacent parts.
- *e*) In a dimension diagram, such as that shown in Fig. 4.1, there should be only one vector for each part. If there are two vectors for any one part then the choice of dimension is in error.

Fig. 4.2 shows a typical dimension diagram for an interference fit condition. The direction of the vector of the dimension condition $(+y_w)$ is represented as if there were no clearance. The *numerical* value of the dimension condition must be negative.

Fig 4.1 Dimensioning diagram using vector system

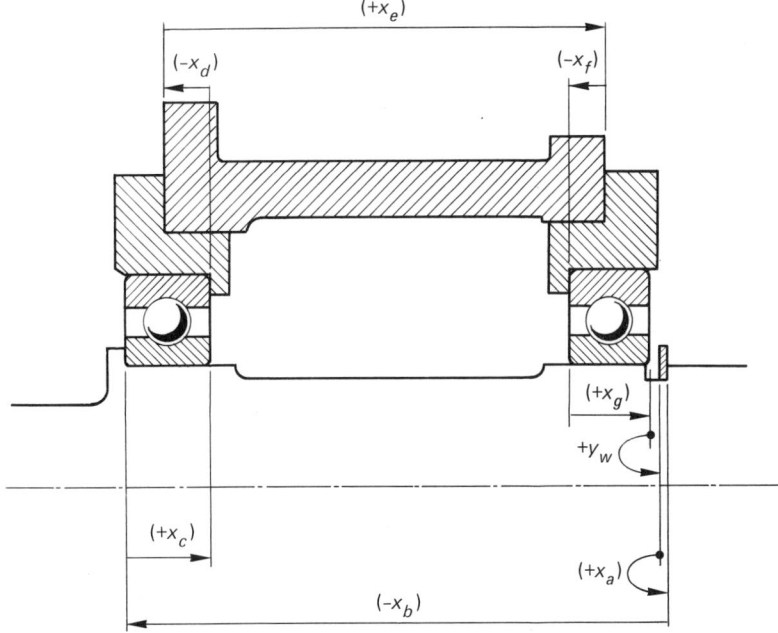

Fig 4.2 Dimension diagram for interference fit

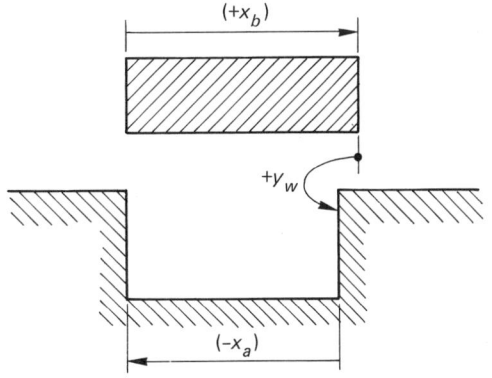

Use of dimension diagrams

a) Ideal for analysing preliminary design studies in order to determine what effect tolerances have on the dimension condition, thereby giving some indication of the dimensional feasibility of the design.

b) They provide the basis for calculating and assigning values to dimensions.

General Equation and Limit Values

The systematic dimensioning method is based on the principle that, when all the vectors are summed together, the result is zero, thus:

$$(+y_w) + \sum (+x) - \sum (-x) = 0$$

where $\sum (+x) =$ sum of rightward vectors

$\qquad \sum (-x) =$ sum of leftward vectors.

It is very simple from this to establish the dimension condition vector $(+y_w)$ by simple transposition.

Note that the positive and negative signs before the x and y values refer only to the direction in which the vector acts.

The following set of equations relate the dimension condition and the dimension vectors:

$$W - w = \sum t \qquad (1)$$

$$W + \sum (+x)_{min} = \sum (-x)_{max} \qquad (2)$$

$$w + \sum (+x)_{max} = \sum (-x)_{min}. \qquad (3)$$

where $w =$ least value of $(+y_w)$

$\qquad W =$ greatest value of $(+y_w)$

$\qquad \sum t =$ sum of all the tolerances for $(-x)$ and $(+x)$ values.

4.2 Dimension Diagrams: a Worked Example

Having set out the basic rules, the following example illustrates how they are applied.

Referring to Fig. 4.1 let

$$w = (+y_w)_{min} = 0.125 \text{ mm}$$

$$W = (+y_w)_{max} = 0.885 \text{ mm}$$

Dimensions x_a, x_c and x_g have limited values since they are dimensions of standard parts. From manufacturers' catalogues the data for these limited dimensions are

Dimension	max	min	$t\ (=\text{max}-\text{min})$
x_a	1.321	1.245	0.076
x_c	13.000	12.873	0.127
x_g	13.000	12.873	0.127

Substituting numerical values into equation (1),

$$0.885 - 0.125 = 0.076 + t_b + 0.127 + t_d + t_e + t_f + 0.127$$

which reduces to

$$0.430 = t_b + t_d + t_e + t_f = \sum t_u \qquad (4)$$

where $\sum t_u$ denotes the sum of tolerances for dimensions which are unrestricted.

The ideal objective is to derive a set of tolerances which allows a designer to allocate them to the mating parts in such a way that the cost of producing them is established. One method of doing this is to apply what are called

Table 4.1

Dimension	d_p	d_m	d_f	d_s	D	t_u
x_b	1.4	2.0	1.3	1.2	5.9	0.121
x_d	1.8	1.6	1.2	0.7	5.3	0.109
x_e	1.2	1.0	1.0	1.2	4.4	0.091
x_f	1.8	1.6	1.2	0.7	5.3	0.109

$$\sum D = 20.9 \quad \sum t_u = 0.430$$

design factors. These factors can be chosen from a designer's own experience of how difficult it is to maintain a tolerance, i.e. a high design factor indicates more difficulty in keeping to a close tolerance. The factors can take account of

a) Type of machining process (d_p).
b) Material used (d_m).
c) The shape (form) of the feature (d_f).
d) The size of the part (d_s).

A chart can then be produced to list these factors relative to the unrestricted dimensions x_b, x_d, x_e, x_f, as in Table 4.1.

The value of D in Table 4.1 is calculated by adding together d_p, d_m, d_f, d_s, thus:

$$D_b = 1.4 + 2.0 + 1.3 + 1.2 = 5.9$$
$$D_d = 1.8 + 1.6 + 1.2 + 0.7 = 5.3$$
$$D_e = 1.2 + 1.0 + 1.0 + 1.2 = 4.4$$
$$D_f = 1.8 + 1.6 + 1.2 + 0.7 = 5.3$$

giving

$$\sum D = 5.9 + 5.3 + 4.4 + 5.3 = 20.9$$

The individual tolerances t_u are then calculated for each unrestricted dimension in the following way:

$$t_u = \frac{D \times \sum t_u}{\sum D}$$

$$t_{ub} = \frac{5.9 \times 0.43}{20.9} = 0.121$$

$$t_{ud} = \frac{5.3 \times 0.43}{20.9} = 0.109$$

$$t_{ue} = \frac{4.4 \times 0.43}{20.9} = 0.091$$

$$t_{uf} = \frac{5.3 \times 0.43}{20.9} = 0.109$$

Table 4.2

Dimension	t	(+x) max	(−x) min	Item
x_a	0.076	1.321		Washer
x_b	0.121		71.120	Shaft
x_c	0.127	13.000		Bearing
x_d	0.109		9.525	End cap
x_e	0.091	$x_{e\ max}$		Housing
x_f	0.109		9.525	End cap
x_g	0.127	13.000		Bearing

$$w = 0.125$$

$$w + \sum (+x)\ max = \sum (-x)\ min$$

$$0.125 + 1.321 + 13.000 + x_{e\ max} + 13.000$$
$$= 71.120 + 9.525 + 9.525$$

$$x_{e\ max} + 27.446 = 90.170$$
$$x_{e\ max} = 62.724\ mm$$

All the values of tolerances have now been established for all required dimensions. From one limit and the tolerance, the other limit can be found.

In order to establish the required dimensions it is necessary to choose values for all unrestricted dimensions except one. Table 4.2 shows how the final dimension $x_{e\ max}$ is found.

Maximum values are tabulated for positive vectors, and minimum values for negative vectors. This results in an $(+x)$ max column and a $(-x)$ min column.

$w = 0.125$ is placed under the $(+x)$ max column and added to these dimensions, just as if the tolerances were being added in the normal way.

The final tolerances for the individual parts are

			max.	min.	
x_a	Washer		1.321	1.245	← 1.321 − 0.076
x_b	Shaft	71.120 + 0.121 →	71.241	71.120	
x_c & x_g	Bearing		13.000	12.873	← 13.000 − 0.127
x_d & x_f	End cap	9.525 + 0.109 →	9.634	9.525	
x_e	Housing		62.724	62.633	← 62.724 − 0.091

By using the above tolerances, the w and W values may be checked to be 0.125/0.885, via the formulae

$$w + \sum (+x)\ max = \sum (-x)\ min \quad \text{and} \quad W + \sum (+x)\ min = \sum (-x)\ max$$

This method allows the designer to maintain all records of important dimensions in a systematic way and to see quickly which parts are affected should the design need to be changed.

Fig 4.3 Pressure fit: shaft and hub

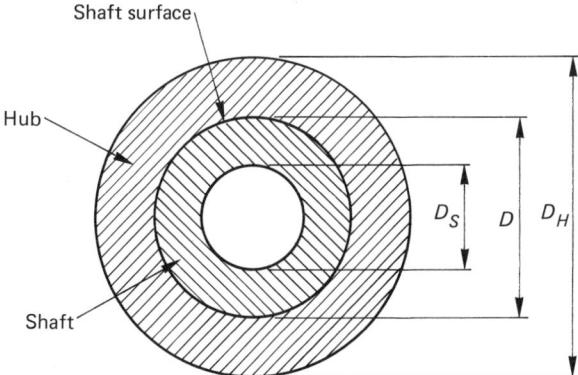

4.3
Pressure Fits

A pressure fit maintains pressure at the shaft surface because of the elastic properties of the materials. The unit pressure P at the interface of shaft and hub (Fig. 4.3) is measured in N/mm², and may be calculated using the Lamé Equation which states

$$P = \frac{i}{D\left[\dfrac{1}{E_H}(\rho_H + \nu_H) + \dfrac{1}{E_S}(\rho_S - \nu_S)\right]} \tag{1}$$

where i = amount of interference (mm)
$É_H, E_S$ = Youngs modulus for hub and shaft respectively (N/mm²).

$$\rho_H = \frac{(D_H/D)^2 + 1}{(D_H/D)^2 - 1} \qquad \rho_S = \frac{1 + (D_S/D)^2}{1 - (D_S/D)^2}$$

ν_H, ν_S = Poisson ratios.
For solid shafts, and hubs which are very large compared with their hole diameter, equation (1) becomes

$$P = \frac{i}{D\left[\dfrac{1}{E_H}(\rho_H + \nu_H) + \dfrac{1}{E_S}(1 - \nu_S)\right]} \tag{2}$$

1 The *torque* which can be transmitted by a pressure fit is given by the formula:

$$T = \tfrac{1}{2}\pi\mu PLD^2$$

where μ = coefficient of friction
P = pressure allowable at shaft/hub interface (N/mm²)
L = length of the joint (mm)
D = shaft diameter (mm).

Table 4.3 Limiting values of ν, E and material stress σ

Material	ν	E (N/mm^2)	σ (N/mm^2)
Cast iron	0.240	131 473	366.7
Cast steel	0.265	200 670	470.5
Alloy steel	0.295	207 590	1384.0
Cast aluminium	0.330	71 964	138.4
Wrought aluminium	0.348	73 348	325.2
Phosphor bronze	0.380	120 000	—

2 *Coefficients of friction μ*

1) Clean unlubricated surfaces	0.3 to 0.5
2) Lubricated surfaces (solid film lubricant)	0.06 to 0.12
3) Lubricated surfaces (not solid film)	0.05 to 0.10

3 If the *stresses* of the shaft and hub are to be within safe limits, then it is necessary to satisfy the following:

Shaft stress $\qquad \sigma_S \geqslant \lambda_S P$
Hub bore stress $\sigma_H \geqslant \lambda_H P$

where

$$\lambda_S = \frac{2}{1 - (D_S/D)^2}$$

$$\lambda_H = \frac{2(D_H/D)^2}{(D_H/D)^2 - 1}$$

Table 4.3 gives typical values for various materials.

4 Force required to separate/assemble the parts is

$$F = \pi \mu P L D$$

where μ, P, L, D are as above.

4.4
Pressure Fits:
Worked Example 1
(Fig. 4.4)

Determine the following given that the shaft fit is an H7/s6 (chart 6).
Coefficient of friction $= 0.3$
  a) Whether the selected fit is suitable for the application.
  b) The force required to assemble the parts.
  c) The minimum torque that can be transmitted by the assembly.
  d) The factor of safety of fit.

Fig 4.4

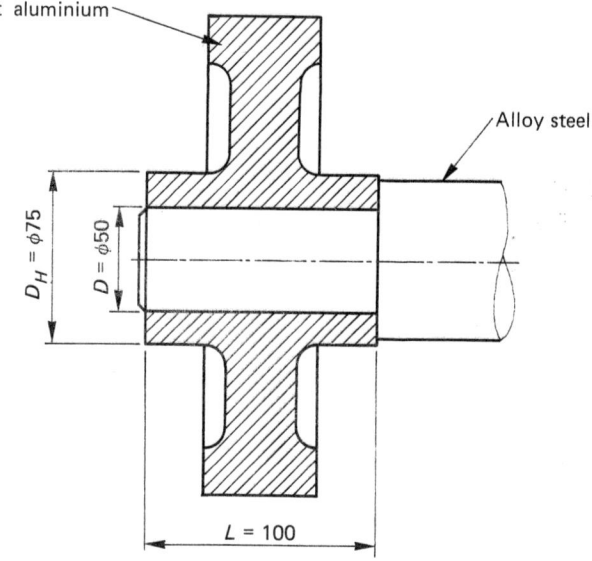

Cast aluminium

Alloy steel

$D_H = \phi 75$

$D = \phi 50$

$L = 100$

From Table 4.3

$$E_H = 71\,964\,\text{N/mm}^2 \qquad E_S = 207\,590\,\text{N/mm}^2$$

$$\nu_H = 0.330 \qquad \nu_S = 0.295$$

$$\sigma_H = 138.4\,\text{N/mm}^2 \qquad \sigma_S = 1384\,\text{N/mm}^2$$

$$\frac{D_H}{D} = \frac{75}{50} = 1.5 \qquad \frac{D_S}{D} = 0$$

$$\rho_S = \frac{1+(0)^2}{1-(0)^2} = 1.0$$

$$\therefore \quad P = \frac{i}{50\left\{\dfrac{1}{71\,964}\left[\left(\dfrac{1.5^2+1}{1.5^2-1}\right)+0.33\right]+\dfrac{1}{207\,590}[1-0.295]\right\}}$$

$$= 453.402i \tag{1}$$

The two inequalities must now be satisfied:

$$\sigma_S \geqslant \lambda_S P \quad \text{and} \quad \sigma_H \geqslant \lambda_H P \tag{2}$$

$$\lambda_S = \frac{2}{1-(0)^2} = 2$$

$$\lambda_H = \frac{2(75/50)^2}{(75/50)^2-1} = 3.6$$

$$\sigma_S = 1384\,\text{N/mm}^2 \qquad \sigma_H = 138.4\,\text{N/mm}^2$$

For hub $\quad 138.4 \geqslant 3.6P$

For shaft $\quad 1384 \geqslant 2P$

For hub, pressure must be equal to or less than $38.44\,\text{N/mm}^2$ (i.e. 138.4/3.6).

For shaft, pressure must be equal to or less than $692\,\text{N/mm}^2$ (i.e. 1384/2).

The least of these must be basis for the design (aluminium hub). Therefore, substitute into $P = 453.402i$.

$$38.44 = 453.402i$$

$$i = 0.0848\,\text{mm}$$

$\therefore$ Maximum allowable interference $= 0.0848$ mm

The H7/s6 fit is

$$
\left.
\begin{array}{l}
\text{HOLE } \varnothing 50 \begin{array}{l} +0.025 \\ -0.000 \end{array} \\[2ex]
\text{SHAFT } \varnothing 50 \begin{array}{l} +0.059 \\ +0.043 \end{array}
\end{array}
\right\}
\begin{array}{l}
0.059 \\
0.018 \text{ mm} \\
\text{interference}
\end{array}
$$

The H7/s6 fit is safe for the application   [Answer a)]

Now the actual maximum tensile stress due to H7/s6 fit must be checked.
Using max. fit of 0.059,
From (1) $P = 453.402 \times 0.059 = 26.751\,\text{N/mm}^2$
From (2) $\sigma_H = 26.751 \times 3.6 = 96.30\,\text{N/mm}^2$

$$\text{Factor of safety} = \frac{138.4}{96.30} = 1.44 \quad [\text{Answer } d)]$$

The force required to assemble the shaft and hub, assuming the coefficient of friction is 0.3, is

$$F = \pi \times 0.3 \times 26.751 \times 100 \times 50$$

$$= 126\,061\,\text{N} \quad [\text{Answer } b)]$$

Note   The worst fit must be taken, and also the actual pressure due to H7/s6, i.e. $P = 26.751\,\text{N/mm}^2$.

The torque which can be transmitted must be based on the minimum interference fit, i.e. 0.018 mm.

$$T = \tfrac{1}{2}\pi \times 0.3 \times (453.402 \times 0.018) \times 100 \times 50^2$$

$$= 961\,473\,\text{N mm} = 961\,\text{N m} \quad [\text{Answer } c)]$$

4.5
Pressure Fits:
Worked Example 2
(Fig. 4.5)

Repeat Example 1 but this time with a 20 mm diameter hole up the centre of the shaft.

$$\frac{D_H}{D} = 1.5 \qquad \frac{D_S}{D} = 0.4$$

$$P = \frac{i}{50\left\{\dfrac{1}{71\,964}\left[\left(\dfrac{1.5^2+1}{1.5^2-1}\right)+0.33\right]+\dfrac{1}{207\,590}\left[\dfrac{1+0.4^2}{1-0.4^2}-0.295\right]\right\}}$$

$$= 435.29i \tag{1}$$

$$\sigma_S \geqslant \lambda_S P \quad \text{and} \quad \sigma_H \geqslant \lambda_H P \tag{2}$$

Fig 4.5

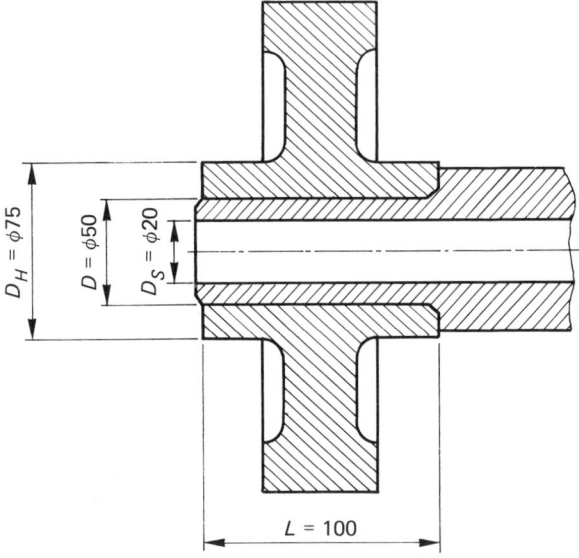

$$\lambda_S = \frac{2}{1-0.4^2} = 2.38$$

$$\lambda_H = \frac{2(75/50)^2}{(75/50)^2 - 1} = 3.6$$

For hub $138.4 \geqslant 3.6P$ $\therefore$ $P = 38.44$ N
For shaft $1384 \geqslant 2.38P$ $\therefore$ $P = 581.5$ N

$$P = 435.29i$$

$$38.44 = 435.29i$$

$$i = 0.088$$

$\therefore$ Maximum allowable interference $= 0.088$ mm
The H7/s6 is safe for the application   [Answer a)]

From (1) $P = 435.29 \times 0.059 = 25.682$ N/mm^2
From (2) $\sigma_H = 25.682 \times 3.6 = 92.46$ N/mm^2

Factor of safety $= \dfrac{138.4}{92.46} = 1.50$ [Answer d)]

Force to assemble shaft and hub is

$$F = \pi \times 0.3 \times 25.682 \times 100 \times 50$$

$$= 121\,024 \text{ N} \text{ [Answer } b)]$$

Torque that can be transmitted is

$$T = \tfrac{1}{2}\pi \times 0.3 \times (435.29 \times 0.018) \times 100 \times 50^2$$

$$= 923\,065 \text{ N mm} = 923 \text{ N m} \text{ [Answer } c)]$$

4.6
Bearing Fits

Bearings usually have interference fits between their outer diameters and housing bores. Allowance must therefore be made for the resulting closure of bearing diameters which would otherwise decrease the required clearance fits for free-running parts, and cause possible seizure.

The following formula may be used for plain bearings (bushes):

$$\Delta_S = \frac{\lambda P D_S}{E_S}$$

where Δ_S = change in diameter at bearing bore (mm)
$\quad P$ = pressure at interface of bearing and housing (N/mm^2)
$\quad D_S$ = bearing nominal bore (mm)
$\quad E_S$ = Young's modulus for the bearing material (N/mm^2)

$$\lambda = \frac{2}{1-(D_S/D)^2}$$

$\quad D$ = bearing nominal outside diameter (mm).

4.7
Bearing Fits:
Worked Example

Fig. 4.6 shows a phosphor bronze plain bearing of outside diameter 60 mm (s6) and required bore tolerance band H7, which is to have an interference fit in a cast iron housing bore of $\varnothing$60 mm (H7) and support a free-running shaft of $\varnothing$50 mm (g6). The housing is a hub with an outside diameter of 80 mm. Determine suitable manufactured limits for the bearing bore if the minimum allowable working clearance between shaft and bearing is 0.009 mm.

Housing bore limits (60 mm H7) = 60.030/60.000 (chart 6)
Bearing outside diameter limits (60 mm s6) = 60.072/60.053 (chart 6)

$\therefore$ Maximum interference $i = 60.072 - 60.000$

$$= 0.072 \text{ mm}$$

$D_H/D = 80/60 = 1.333$

$D_S/D = 50/60 = 0.8333$

$E_H = 131\,473 \text{ N/mm}^2 \qquad E_S = 120\,000 \text{ N/mm}^2$

$\nu_H = 0.24 \qquad \nu_S = 0.38 \qquad$ (Table 4.3)

Using the Lamé equation:

$$P = \frac{0.072}{60\left[\dfrac{1}{131\,473}\left(\dfrac{1.333^2+1}{1.333^2-1}+0.24\right)+\dfrac{1}{120\,000}\left(\dfrac{1+0.8333^2}{1-0.8333^2}-0.38\right)\right]}$$

$$= 16.66 \text{ N/mm}^2$$

$$\lambda = \frac{2}{1-0.8333^2} \qquad \text{(see section 4.6)}$$

$$= 6.544$$

Fig 4.6

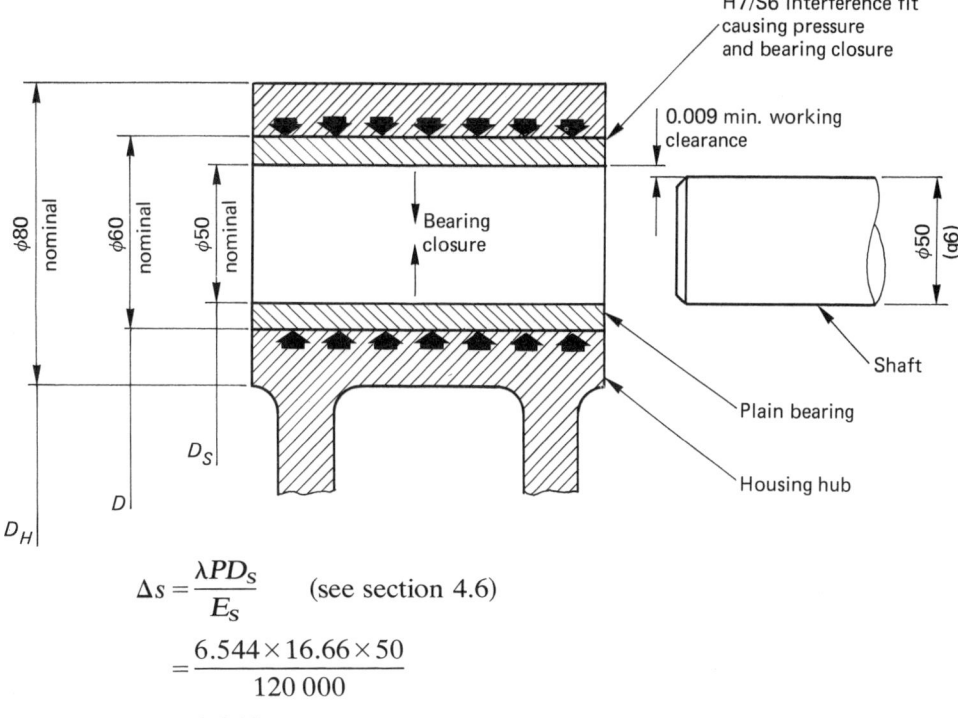

$$\Delta s = \frac{\lambda P D_S}{E_S} \qquad \text{(see section 4.6)}$$

$$= \frac{6.544 \times 16.66 \times 50}{120\,000}$$

$$= 0.046 \text{ mm}$$

Therefore, bearing bore closes in by 0.046 mm on assembly. Initial clearance between bearing and shaft must be at least

$$0.046 + 0.009 = 0.055 \text{ mm}$$

Upper limit of shaft diameter (50 g6) = 49.991 mm   (chart 6)

∴   Lower limit of bearing bore must be at least

$$49.991 + 0.055 = 50.046 \text{ mm}$$

Bearing bore tolerance band (H7, range 40–50) = 0.025 mm   (chart 6)

∴   Minimum allowable upper limit of bearing bore is

$$50.046 + 0.025 = 50.071 \text{ mm}$$

∴   Suitable manufactured bearing bore limits are

50.071/50.046 mm   [Answer]

**4.8
Summary**

In assembling shafts and hubs with an interference fit, it is essential that there exists a force which is exerted between the mating surfaces. This in turn will deform the material, and thus, in the case of a thin-walled section for example, as in a plain bearing, will mean that the inner surface will decrease in size. Stresses will also exist because of these fits and they must be limited to account for the different materials used. The foregoing sections on pressure fits and bearing fits have been included to show that a designer must be careful in selecting fits having interference characteristics, even though they are listed in British Standards.

The method of assembling a shaft and hub having an interference fit is to heat the hub prior to assembly, or freeze the shaft to make the material shrink. This will ensure that the parts will assemble without scruffing the surfaces excessively.

In order to have reasonable tolerances for interchangeability it is sometimes necessary to stress the members beyond the elastic limit of the material. This is possible only with the materials that deform in the plastic range when the elastic limit of the material is exceeded. This technique must be used with care since variations exist in different batches of material having the same basic specification.

Problems

4.1 Find the dimensions and their tolerances for the assembly shown in Fig. 4.7.

Let $w = 0.05$ mm and $W = 0.52$ mm.

From manufacturers' catalogues the data for limited dimensions x_a and x_c are as follows:

Dimension	x_{max}	x_{min}	t
x_a	10.00	9.88	0.12
x_c	10.00	9.88	0.12

$-x_{b\ min} = 74.12$ mm

Design factors

	Process	Material	Shape	Size
x_b	1.4	2.0	1.3	1.2
x_d	1.2	1.0	1.0	1.2

[*Ans.* $x_a = x_c = 10.000/9.880$
$x_b = 74.252/74.120$, $x_d = 54.070/53.972$]

4.2 Find the dimensions and their tolerances for the assembly shown in Fig. 4.8.

Let $w = 0.07$ mm and $W = 0.54$ mm.

From manufacturers' catalogues the data for limited dimensions x_c and x_b are as follows:

Dimension	x_{max}	x_{min}	t
x_a	1.300	1.225	0.075
x_b	12.000	11.880	0.120
x_c	12.000	11.880	0.120

$-x_{e\ min} = 74.5$ mm

Fig 4.7

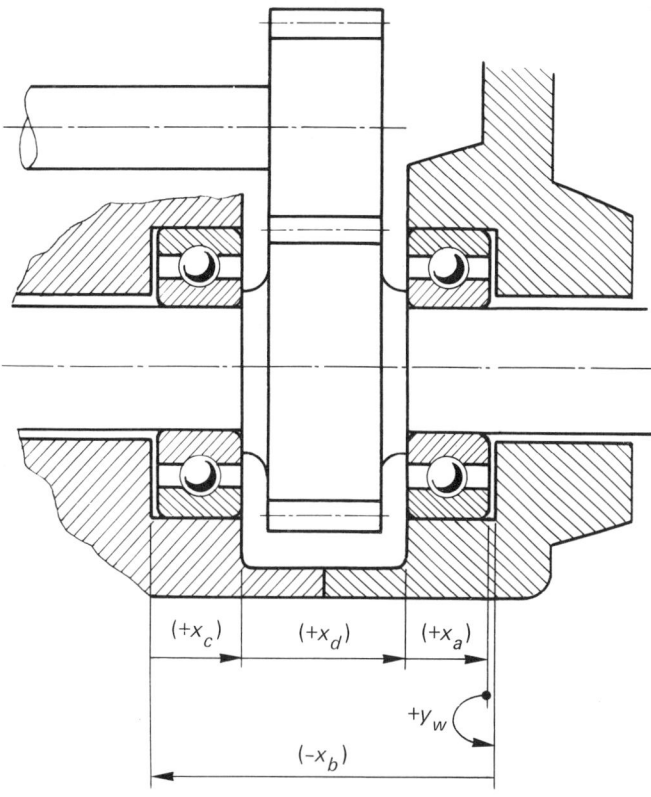

Fig 4.8

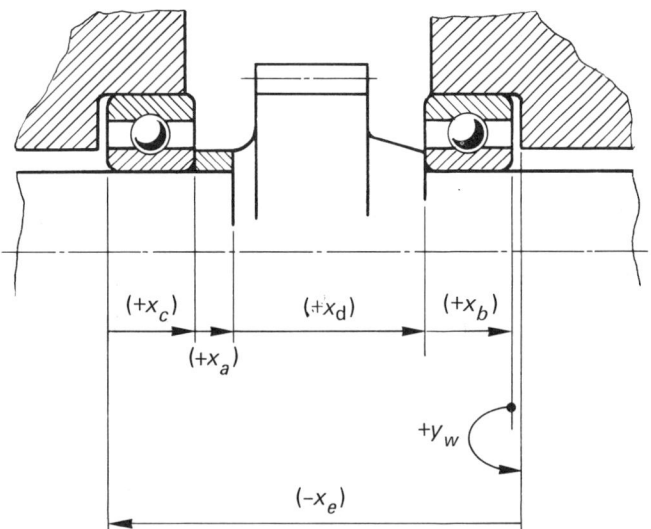

Fig 4.9

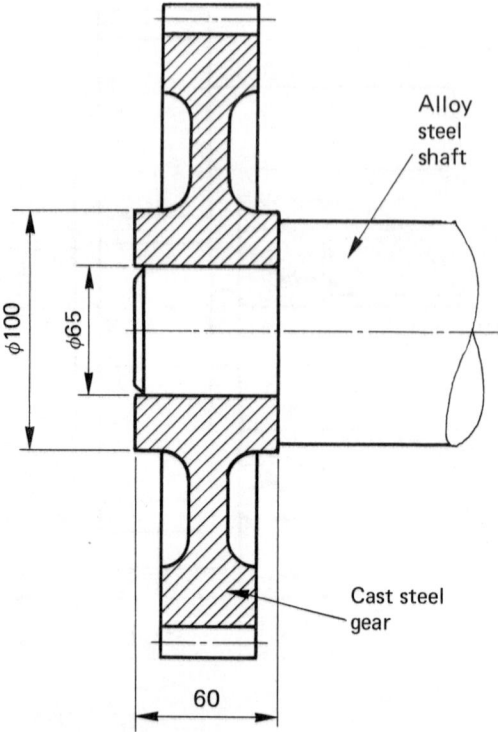

Design Factors

	Process	Material	Shape	Size
x_d	1.2	1.5	1.1	1.2
x_e	1.4	1.0	1.0	1.2

[*Ans.* $x_b = x_c = 12.000/11.880,$ $x_e = 74.574/74.500,$
$x_d = 49.130/49.049,$ $x_a = 1.300/1.225$]

4.3 *a*) Determine the minimum interference fit required between a hub and shaft assembly of the proportions shown in Fig. 4.9 such that it can transmit a torque of 4216 N m. ($\mu = 0.3$.)
b) Find the maximum allowable interference based on maximum stress limit from Table 4.3.

   [*Ans. a*) 0.039, *b*) 0.150]

4.4 A phosphor bronze plain bearing of outside diameter 30 mm (s6) and bore tolerance band H7 is to have an interference fit with a cast steel hub of bore $\varnothing$30 mm (H7) and outside diameter 48 mm. The bearing is to support a free-running shaft of $\varnothing$24 mm (g6). Determine suitable manufactured limits for the bearing bore if the minimum allowable working clearance between shaft and bearing is 0.007 mm.
   [*Ans.* 24.058/24.037 mm]

Bibliography
Earlwood T. Fortini *Dimensioning for Interchangeable Manufacture* (Industrial Press, New York)

5 Analysis and Calculation

Chapter 2 of *Engineering Design for Technicians* devoted some length to the calculation of design dimensions with particular emphasis on stress conditions in screws, bolts, pins and keys. This chapter may be considered as a continuation of those exercises with more emphasis on shaft design, component sections, structural dimensions, bearing selection and gear design. As in the previous book, charts of materials properties are supplied at the back of the book, to aid the design process.

5.1 Factors of Safety

Quoted values of tensile, compressive and shear strengths of various materials are obtained under ideal static conditions. In service, however, machines are subjected to influences which are difficult to allow for in analytical calculations. These may include:

a) Additional shock loading.

b) Slight inaccuracies of manufacture or installation.

c) Environmental conditions.

d) High risk to human life or damage to other components.

e) Varying frequencies and durations of operation.

The designer must therefore apply a factor of safety to the quoted strength figure based on expert knowledge of the service conditions which the design will encounter. For example, for tensile strengths,

$$\text{Allowable design stress} = \frac{\text{Tensile Strength}}{\text{Factor of Safety}}$$

In some design applications it is more suitable to apply the factor of safety to the fatigue limit, elastic limit or yield stress of the material.

5.2 Shaft Design

Shafts may fail due to torsional loading, to bending, and, to a lesser extent, to axial loading, either separately or in combination.

Keyways in Shafts

A keyway in a shaft will cause a localised weakness which must be accounted for in calculations. For general applications, it may be assumed that, for shafts having standard keyways, the allowable design stress must be about 75% of the allowable stress without a keyway:

Allowable stress with keyways
$= 0.75 \times$ Allowable stress without keyways

For applications of pronounced fatigue loading, an alternative method involving stress concentration factors (K_t) values is recommended, as described in pages 86–91 of this chapter.

Fig 5.1

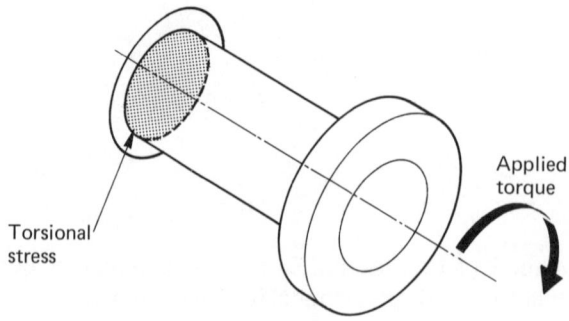

Torsional stress

Applied torque

Torsion in Shafts (Fig. 5.1)

Torsion is a shear stress condition caused by the twisting effect of torque loadings. Shaft torsion is caused mainly by the direct coupling torque of motors and engines, or by tangential forces on gears, belts and chain drives.

Formulae associated with torsion loading include:

1) $P = T\omega$

2) $J = \dfrac{\pi D^4}{32}$

3) $\dfrac{T}{J} = \dfrac{\tau}{r}$

where P = power (watts)
T = torque (N m or N mm)
ω = angular velocity (rad/sec)
J = polar 2nd moment of area for a circle (mm⁴)
D = shaft diameter (mm)
r = shaft radius (mm)
τ = shear stress (N/mm²)

Note: Formula 1) requires the use of the SI torque unit (N m). In formulae 2) and 3), millimetre units may be used, i.e. mm, mm⁴, N mm, N/mm².

Example 5.1   A light-duty shaft is directly coupled to a 7 kW electric motor which rotates at 1500 rev/min. The connection is via a flexible coupling fixed with standard keys. For a required safety factor of 5, calculate the minimum allowable diameter of shaft.

Solution   For this light-duty application, 070M20 (EN3) steel can be used for the shaft.

Shear strength of 070M20 (EN3) = 270 N/mm²   (see chart 2, p.218).

Allowable stress (without keyways)

$= \dfrac{270}{5} = 54 \text{ N/mm}^2$

Allowable stress (with keyways)
$$= 0.75 \times 54 = 40.5 \text{ N/mm}^2$$

Angular velocity $\omega = \dfrac{2\pi \times 1500}{60}$

$$= 157.1 \text{ rad/sec}$$

$$P = T\omega$$

$$\therefore \quad T = P/\omega = 7000/157.1 = 44.56 \text{ N m} = 44\,560 \text{ N mm}$$

$$J = \pi D^4/32$$

$$T/J = 40.5/r$$

$$\therefore \quad \frac{44\,560}{\pi D^4/32} = \frac{40.5}{D/2} \qquad \frac{44\,560}{\pi D^3/32} = \frac{40.5}{1/2}$$

$$\therefore \quad D = \sqrt[3]{\left[\frac{\frac{1}{2} \times 44\,560 \times 32}{\pi \times 40.5}\right]} = 17.76 \text{ mm}$$

Chosen shaft diameter $= 18$ mm [Answer]

Fig 5.2

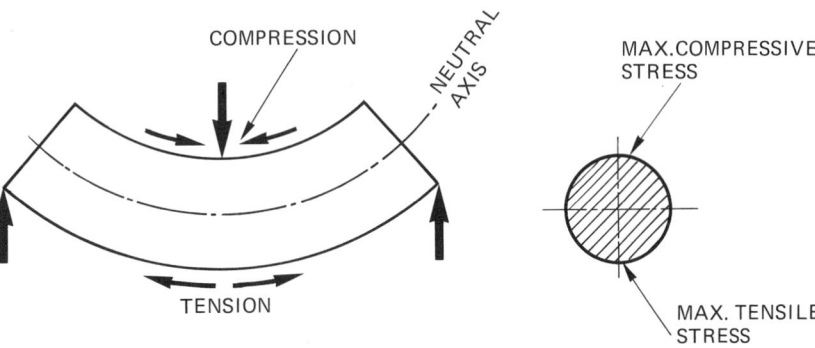

COMPRESSION NEUTRAL AXIS MAX.COMPRESSIVE STRESS

TENSION MAX. TENSILE STRESS

Bending of Shafts (Fig. 5.2)

Bending stresses occur in shafts as a result of applied weights such as pulleys and gears, the shaft weight itself, belt or chain tensions, and the normal force of gear teeth. Bending is a mixture of tensile and compressive forces which increase with the distance either side of the neutral axis. This axis always passes through the centroid of the cross-section. Relevant formula for shaft bending include:

$$I = \frac{\pi D^4}{64} \quad \text{and} \quad \frac{M}{I} = \frac{\sigma}{y}$$

where $I = $ 2nd moment of area for circle about neutral axis (mm^4)
  $D = $ shaft diameter (mm)
  $M = $ maximum bending moment (N mm)
  $\sigma = $ maximum bending stress (N/mm^2)
  $y = $ distance of outside edge to neutral axis (mm)

Fig 5.3

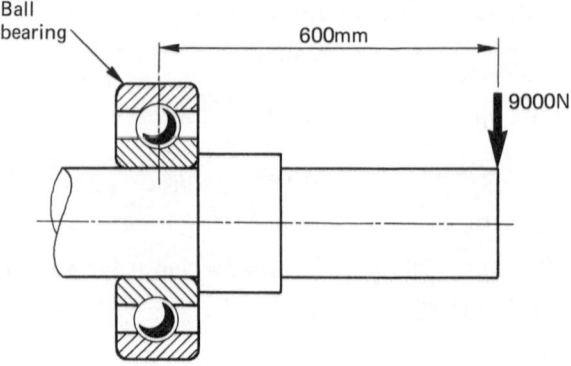

Example 5.2 Fig. 5.3 shows a shaft assembly which supports an overhung load of 9000 N. If the shaft is made of 080M40 (EN8) steel and a safety factor of 10 is required, determine the minimum allowable shaft diameter.

Solution In these problems it is usually assumed that the reaction to the load acts at the centre of the supporting bearing width, and thus in this case the maximum cantilever bending moment is at the bearing width centre.

Tensile/compressive strength of 080M40 (EN8)

$= 620 \text{ N/mm}^2$ (see chart 2)

∴ Allowable bending stress $\sigma = 620/10 = 62 \text{ N/mm}^2$

Max. bending moment $M = 9000 \times 600 = 5.4 \times 10^6$ N mm

Since $I = \pi D^4/64$ and $M/I = \sigma/y$

$$\frac{5.4 \times 10^6}{\pi D^4/64} = \frac{62}{D/2} \quad \text{i.e.} \quad \frac{5.4 \times 10^6}{\pi D^3/64} = \frac{62}{1/2}$$

$$\therefore \quad D = \sqrt[3]{\left[\frac{5.4 \times 10^6 \times \frac{1}{2} \times 64}{\pi \times 62}\right]} = 96.09 \text{ mm}$$

Chosen shaft diameter $= 100$ mm   [Answer]
(This would conform to a convenient bearing bore size.)

Combined Bending and Torsion

The previous examples were concerned with torsion or bending separately. In most cases, however, the applied loading is a combination of both. The important formulae of use here are

$$q = \sqrt{\left[\left(\frac{\sigma}{2}\right)^2 + \tau^2\right]}$$

$$T_e = \sqrt{[M^2 + T^2]}$$

$$\frac{T_e}{J} = \frac{q}{r}$$

$M =$ maximum bending moment (N mm)
$T =$ applied torque (N mm)
$\sigma =$ bending stress (N/mm^2)
$\tau =$ shear stress (N/mm^2)
$q =$ combined bending and shear stress (N/mm^2)
$T_e =$ equivalent torque due to bending and torsion (N mm)
$J =$ polar 2nd moment of area (mm^4)
$r =$ shaft radius (mm)

Gear Loadings on Shafts

Chapter 5 of *Engineering Design for Technicians* listed and described various types of gear system. The mechanical loading of some gear-types may be varied and complex. In this section the discussion is confined to simple applications involving spur gears and helical gears. Information on gear proportions is supplied in BS436 Parts I and II and is summarized on page 95 of this chapter.

1 In *spur gears*, the line of action of tooth loading is governed by its pressure angle of involute form. It is convenient to split the tooth load into two components (Fig. 5.4):

   i) The tangential force F_t
   ii) The tooth separating force F_s acting towards the centre of the shaft.

Tooth load formulae for spur gears are

$$F_t = \frac{T}{R_{PC}} \qquad \text{and} \qquad F_s = F_t \times \tan \psi$$

where F_t = tangential force (N)
      T = torque (N mm)
   R_{PC} = pitch circle radius PCR (mm)
     F_s = tooth separating force (N)
     ψ = pressure angle.

Fig 5.4

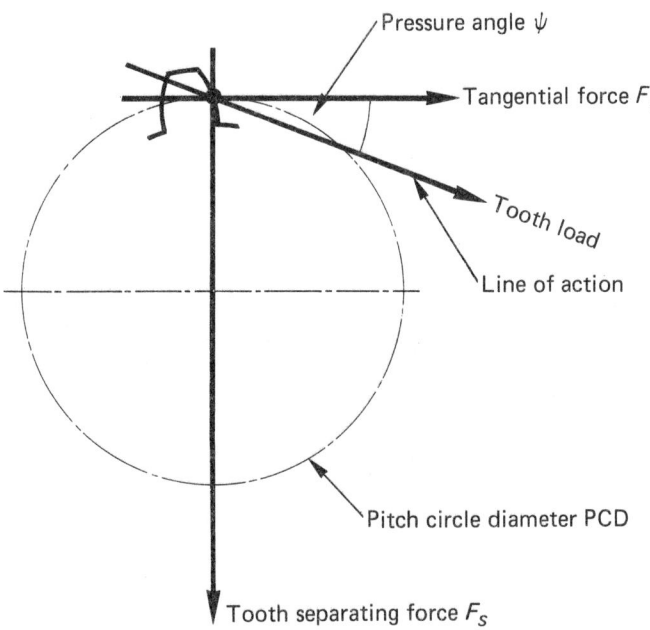

Fig 5.5

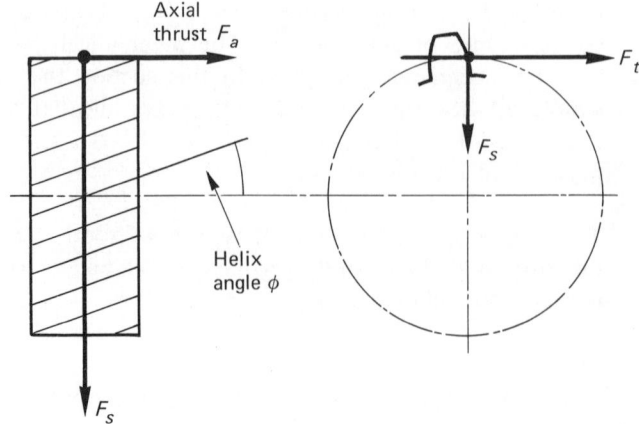

2 With *helical gears*, the supporting shaft and bearings are subjected to an axial thrust force F_a in addition to the tangential and tooth separating forces (Fig. 5.5). The value of the axial thrust is affected by the helix angle of tooth. Although axial thrust can cause problems with slender shafts, it is not generally considered to be a serious problem in shaft design and is far more important in bearing selection as described later in this chapter.

Tooth load formula for helical gears are given as:

$$F_t = \frac{T}{R_{PC}}$$

$$F_s = \frac{F_t \times \tan \psi}{\cos \phi}$$

$$F_a = F_t \times \tan \phi$$

where F_a = axial thrust force (N)
ϕ = helix angle.

5.3
Fatigue Loading on Shafts

Fatigue failure occurs when an engineering component is subjected to a varying load which conforms to a regular cyclic pattern, e.g. an alternating load caused by a vibration such as a fluctuating bending load caused by a shaft rotation.

Fatigue loading has the effect of gradually enlarging small cracks in the material surface. The effect is due to a concentration of stress intensity (or "stress-raiser") at the roots of the cracks. The size of the crack, and the extent of the stress concentration, continue to increase until the average stress along the cross-section is high enough to cause a sudden fracture.

Fatigue Limit of Materials

The fatigue limit of a particular material is the maximum stress which can be applied for an unlimited number of times without causing failure. If the applied stress exceeds the fatigue limit, then the component can only be expected to last for a limited number of cycles before failure occurs. If the applied stress is kept below the fatigue limit, the component could be expected to have an unlimited life.

Fig 5.6

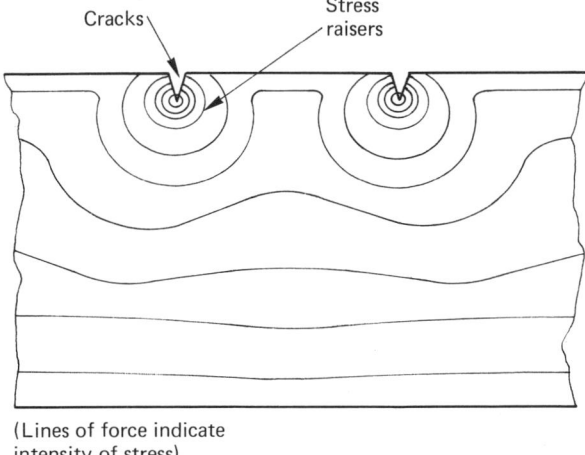

(Lines of force indicate
intensity of stress)

Fig 5.7 Typical
example of *S/N*
diagram

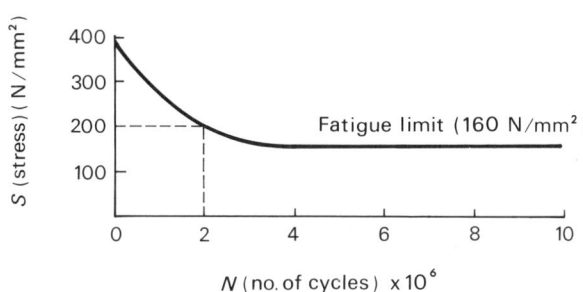

Fatigue limits may be indicated on *S/N* Diagrams which plot applied stress against number of cycles expected life (Fig. 5.7).

Using the *S/N* diagram in Fig. 5.7, it will be seen that, if a stress of, say, 200 N/mm^2, is applied, a component will only be expected to last for 2×10^6 cycles before fracture. If, however, the applied stress is below the fatigue limit of 160 N/mm^2, an unlimited life may be expected (assuming the required safety factors are applied.)

Unfortunately, fatigue limits are not often supplied in specifications for materials. Thus a rough estimate is often taken as a direct ratio to the static strength figure. For steels it may be generally assumed that the fatigue limit will be at least 40% of the static strength quoted. For steels,

Bending fatigue limit $= 0.4 \times$ tensile strength

Torsional fatigue limit $= 0.4 \times$ shear strength

(or about $0.2 \times$ tensile strength)

Table 5.1 Approximate stress concentration factors K_t (as given by Peterson)

	K_t (bending)			K_t (torsion)		
	r/D	D/d	K_t	r/D	D/d	K_t
	0.01	1.1	1.5	0.01	1.1	2
	0.05	1.1	1.2	0.05	1.1	1.5
	0.05	1.2	1.2	0.05	1.2	1.6
	0.1	1.1	1.15	0.1	1.1	1.3
	0.1	1.2	1.2	0.1	1.2	1.35
	0.1	2	1.2	0.1	2	1.4
	0.3	1.2	1.1	0.3	1.2	1.15
	0.3	2	1.1	0.3	2	1.2
	2	–	1.04	2	–	1.04
	4	–	1.02	4	–	1.02
	8	–	1.04	8	–	1.01
Standard keyway with $r/D = 0.02$ and semi-circular ends	0.02	–	1.6 (surface)	0.02	–	3.4 (fillet)

Stress Concentration Factors

The fatigue limit figure for a certain material assumes that the component has a fine surface texture and a uniform cross-section. The allowable fatigue stress is in fact greatly affected by additional stress-raisers caused by

a) Sharp corners in profile (i.e. small fillet radii).
b) Abrupt changes in cross-section.
c) Poor surface finish.

The shape of the finished component may be accounted for by applying a stress concentration factor K_t to the fatigue limit. (This is in addition to the required factor of safety.) Thus, for fatigue-loaded components with stress concentrations,

$$\text{Allowable stress} = \frac{\text{Fatigue limit}}{\text{Safety factor} \times \text{Stress concentration factor } K_t}$$

Some approximate values of stress concentration factors are supplied in Table 5.1. These have been compiled from research undertaken by R. E. Peterson.

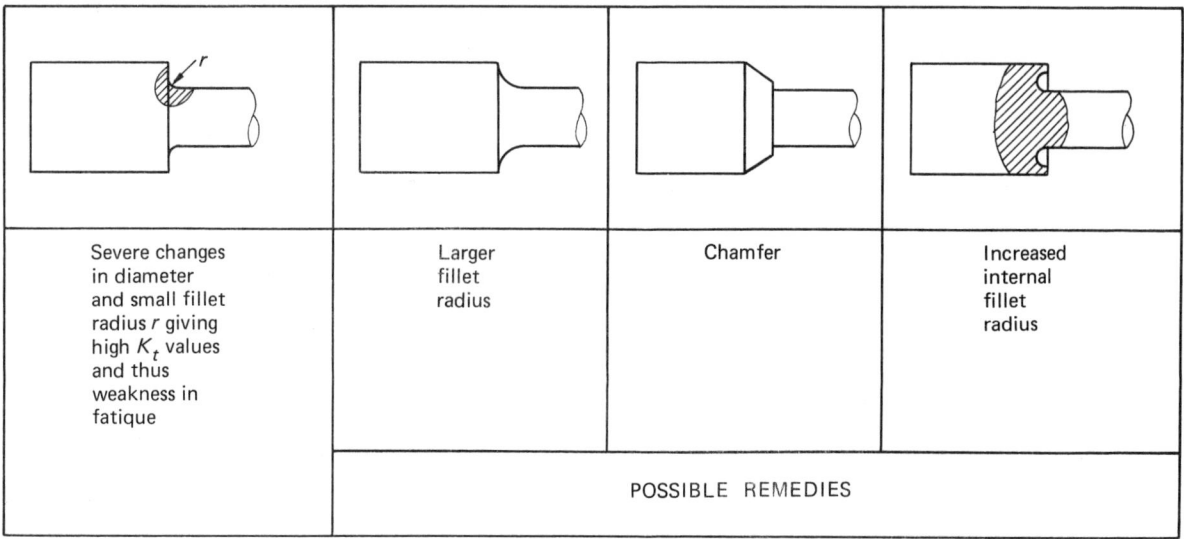

Severe changes in diameter and small fillet radius r giving high K_t values and thus weakness in fatigue	Larger fillet radius	Chamfer	Increased internal fillet radius
	POSSIBLE REMEDIES		

Fig 5.8 **Fig 5.9**

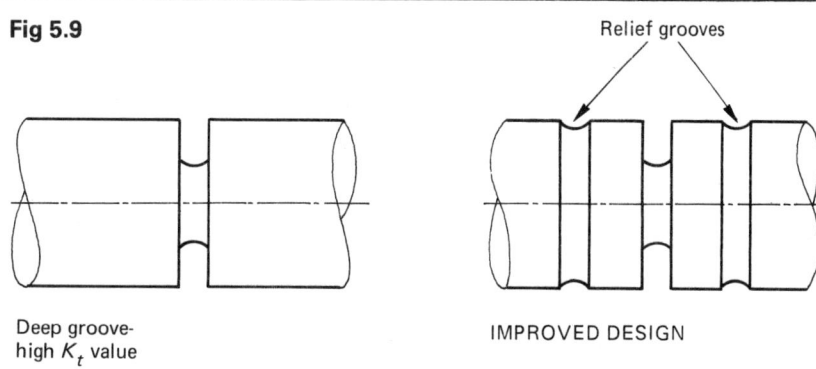

Deep groove-
high K_t value

IMPROVED DESIGN

Relief grooves

Avoiding Fatigue Failure in Shaft Design

This may be tackled by reducing the K_t value to a minimum in the design shape. Fig. 5.8 shows possible remedies for a design likely to fail in fatigue. Deep grooves with small radii will give a high K_t value. This may be overcome by using relief grooves, as shown in Fig. 5.9.

5.4 Keyways Subjected to Fatigue

A general stress factor for keyways was given at the beginning of section 5.2. In cases of more pronounced fatigue loading however, a more comprehensive procedure involving K_t values must be used. The main factors affecting the fatigue strength of standard keyways are the size of fillet radius r and the method of manufacture.

The fillet radii listed for British Standard Keyways are satisfactory in general applications, but would give very high K_t values under pronounced fatigue loading. Data from Peterson suggests larger fillet radii giving an average r/D ratio of just over 0.02 (as shown in Table 5.1).

Keyway fatigue failure may start either at the surface of the shaft (as is more common with bending fatigue) or at the fillet radius (as is more common with torsion fatigue). The K_t values for bending and torsion in Table 5.1 are given on this assumption.

Fig 5.10

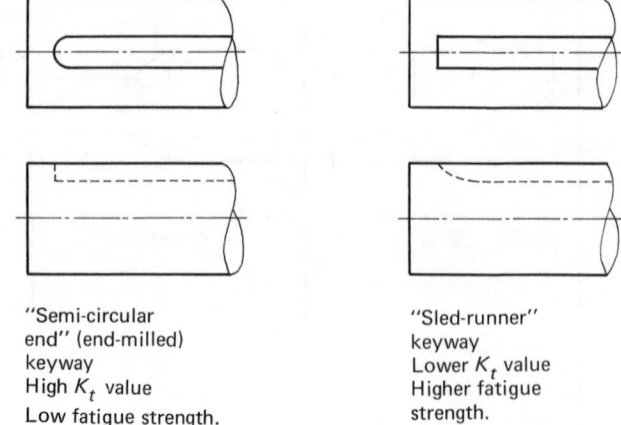

"Semi-circular
end" (end-milled)
keyway
High K_t value
Low fatigue strength.

"Sled-runner"
keyway
Lower K_t value
Higher fatigue
strength.

Fig 5.11

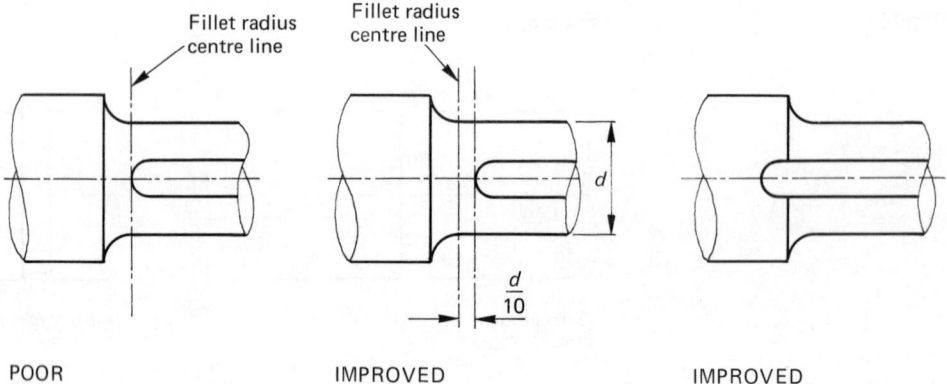

POOR   IMPROVED   IMPROVED

Table 5.1 also assumes the use of the most common type of keyway, i.e. "semi-circular ended", or "mill ended" (Fig. 5.10). This type may only be produced using end-mill which leaves circular tool marks. The K_t value could be considerably reduced using a "sled-runner" type keyway which leaves vastly superior longitudinal tool marks.

Where a semi-circular-ended keyway is close to a shaft shoulder, a high K_t value exists if the keyway ends where the shoulder fillet radius begins. To avoid this, the keyway should either be positioned some distance from the shoulder fillet (about 1/10 of the smaller diameter) or be cut into the shoulder as shown in Fig. 5.11.

Example 5.3 A shouldered shaft made of 070M20 (EN3) steel is required to sustain a maximum bending moment of 4500 N m and torque of 3500 N m under cyclic loading. With a required safety factor of 2 on the fatigue limit, determine suitable dimensions for the shaft if the shoulder is used as a bearing location.

Solution Proportions decided $r/D = 0.01$ $D/d = 1.1$ (by inspection of bearing catalogue)

Tensile strength of 070M20 (EN3) = 460 N/mm^2 (see chart 2)

Bending fatigue limit = 460×0.4

$$= 184 \text{ N/mm}^2 \quad \text{(see p. 85)}$$

From Table 5.1 K_t (bending) = 1.5

Allowable bending stress $\sigma = \dfrac{\text{Fatigue limit}}{\text{Safety factor} \times K_t}$

$$= \frac{184}{2 \times 1.5} = 61.33 \text{ N/mm}^2$$

Shear strength of 070M20 (EN3) = 270 N/mm^2 (see chart 2)

Torsional fatigue limit = 270×0.4

$$= 108 \text{ N/mm}^2 \quad \text{(see p. 85)}$$

From Table 5.1 K_t (torsion) = 2

Allowable torsional stress $\tau = \dfrac{\text{Fatigue limit}}{\text{Safety factor} \times K_t}$

$$= \frac{108}{2 \times 2} = 27 \text{ N/mm}^2$$

Combined stress $q = \sqrt{[(\sigma/2)^2 + \tau^2]}$

$$= \sqrt{[(61.33/2)^2 + 27^2]}$$
$$= 40.86 \text{ N/mm}^2$$

Equivalent torque is

$$T_e = \sqrt{[M^2 + T^2]}$$
$$= \sqrt{[(4.5 \times 10^6)^2 + (3.5 \times 10^6)^2]}$$
$$= 5.7 \times 10^6 \text{ N mm}$$

$$J = \pi d^4/32$$

$$T_e/J = q/r$$

where r = smaller shaft radius (in this formula)

$$\therefore \quad \frac{5.7 \times 10^6}{\pi d^4/32} = \frac{40.86}{d/2} \quad \frac{5.7 \times 10^6}{\pi d^3/32} = \frac{40.86}{1/2}$$

$$d = \sqrt[3]{\left[\frac{5.7 \times 10^6 \times \frac{1}{2} \times 32}{\pi \times 40.86}\right]} = 89.23 \text{ mm} \quad \text{(say 90 mm)}$$

$$\therefore \quad D = 1.1 \times 90 = 99 \text{ mm (say 96 mm)}$$

Fillet radius $r = 0.01 \times 96 = 0.96$ mm (say 1 mm)

Chosen shaft dimensions are as shown in Fig. 5.12. These dimensions would have to be checked against the requirements for the selected bearing. However, they would be suitable for the majority of 90 mm bore bearings.

Fig 5.12

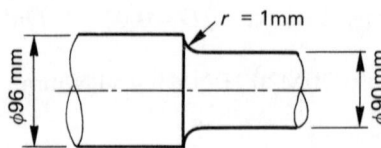

Example 5.4 A keyed shaft made from 08040 (EN8) steel is required to transmit a cyclic torque of 4000 N m with a required safety factor of 3 on fatigue limit. If the keyway is to be close to a shoulder, determine suitable dimensions for the shaft.

Solution Because of manufacturing requirements, an end-milled keyway is chosen.

$$\text{Shear strength of 080M40 (EN8)} = 370 \text{ N/mm}^2 \quad \text{(see chart 2)}$$

$$\text{Torsional fatigue limit} = 0.4 \times 370$$

$$= 148 \text{ N/mm}^2 \quad \text{(see p. 85)}$$

From Table 5.1 K_t (torsion) $= 3.4$

$$\text{Allowable stress } \tau = \frac{\text{Fatigue limit}}{\text{Safety factor} \times K_t}$$

$$= \frac{148}{3 \times 3.4} = 14.5 \text{ N/mm}^2$$

$$J = \pi d^4 / 32$$

$$\frac{T}{J} = \frac{\tau}{r}$$

where r = smaller shaft radius in this formula.

$$\therefore \quad \frac{4 \times 10^6}{\pi d^4 / 32} = \frac{14.5}{d/2}$$

$$\frac{4 \times 10^6}{\pi d^3 / 32} = \frac{14.5}{1/2}$$

$$d = \sqrt[3]{\left[\frac{4 \times 10^6 \times \frac{1}{2} \times 32}{\pi \times 14.5} \right]} = 112 \text{ mm} \quad \text{(say 115 mm)}$$

If the shaft is not to fail at the shoulder fillet, the K_t value here must be no greater than the 3.4 value at the keyway.

Inspection of Table 5.1 reveals that all listed K_t values for a shouldered shaft in torsion are less than 3.4, and so any quoted proportions would be suitable.

Now, chosen proportions are

$$r/D = 0.01 \qquad D/d = 1.1 \qquad (K_t = 2)$$
$$D = 1.1 \times 115 = 126.5 \quad \text{(say 125 mm)}$$
$$r = 0.01 \times 125 = 1.25 \text{ mm} \quad \text{(say 1.5 mm)}$$

Distance of keyway end from shoulder is

$$1.5 + (\tfrac{1}{10} \times 115) = 13 \text{ mm}$$

Chosen shaft dimensions are as shown in Fig. 5.13.

Fig 5.13

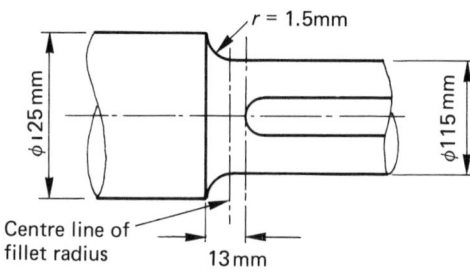

$r = 1.5$mm

ϕ125 mm

ϕ115 mm

Centre line of fillet radius

13 mm

5.5
Bearing Selection

Some of the more common types of anti-friction bearing were listed and discussed in Chapter 7 of *Engineering Design for Technicians*.

Load calculations can be very complex and thus only the simplest types of bearing have been considered. For further reference, the authors strongly recommend the use of comprehensive information supplied in several manufacturer's catalogues. Also, it should be realised that the selection of a bearing for loading capacity will depend, not only upon radial and axial forces, but also on speed of rotation, lubrication, axial alignment, and frequency of use. In particular, the loading capacity of any bearing diminishes rapidly with increase of speed, and thus a speed factor is always applied to the loading calculation.

The formula of most use in bearing selection gives an equivalent load of axial and radial forces:

$$P = XF_r + YF_a$$

where P = equivalent dynamic load (N)
F_r = radial load (N)
X = radial factor of bearing
F_a = axial load (N)
Y = axial factor of bearing.

Bearing Life

Bearing manufacturers often recommend that selection be carried out on the basis of an expected "L10" probability lifetime for a particular loading and speed. An L10 lifetime indicates that no more than 10% of bearings are likely to fail within the stated lifetime (i.e. the bearing has a 90% probability of lasting the stated lifetime). Required bearing lifetimes will vary with different applications, but, for example, one common requirement is 5000 hours (L10).

The recognised calculation procedures are outlined by the International Standards Organisation (I.S.O.) in I.S.O. 281 which states that

1) The life of a ball bearing is inversely proportional to $(\text{Load})^3$
2) The life of a roller bearing is inversely proportional to $(\text{Load})^{10/3}$

The expected life of any bearing under a particular loading condition, will depend on its basic dynamic load rating C. The C value of a bearing is supplied in the bearing manufacturer's catalogue.

The expected lifetime for *ball bearings* may be found from the formula:

$$L = \frac{(C/P)^3 \times 10^6}{60N}$$

where L = expected L10 lifetime (hours)
C = basic dynamic load rating (N)
P = equivalent dynamic load (N)
N = rev/min

The expected lifetime for *roller bearings* may be found from

$$L = \frac{(C/P)^{10/3} \times 10^6}{60N}$$

Example 5.5 A catalogue quotes the following values for a ball bearing:
Basic load rating $C = 60\,000$ N
X factor = 0.56
Y factor = 1.4

If the bearing is subjected to a radial load of 10 000 N, and an axial load of 7500 N, calculate the expected L10 lifetime for a shaft speed of 120 rev/min.

Solution Equivalent dynamic load is

$$P = XF_r + YF_a = (0.56 \times 10\,000) + (1.4 \times 7500) = 16\,100 \text{ N}$$

$$L = \frac{(C/P)^3 \times 10^6}{60N} = \frac{(60\,000/16\,100)^3 \times 10^6}{60 \times 120} = 7189$$

Bearing life is 7189 hours (L10).

Typical Bearing Arrangements

1 *Two Ball Bearings* Ball bearings are designed to accommodate medium duty radial and axial loads. The arrangement shown in Fig. 5.14 is located axially at bearing A. For central loading, the radial load would be shared equally by both bearings, but the axial load would be taken by bearing A only. Bearing B is free to slide on its outer race due to the small clearance

Fig 5.14 Common arrangement for two ball bearings

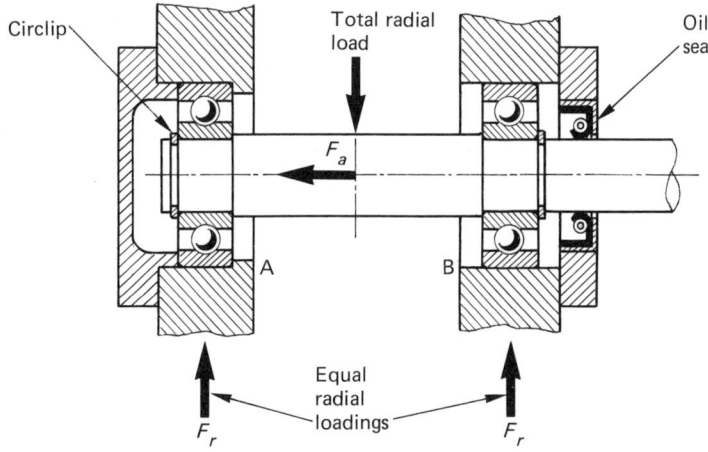

Fig 5.15 Common arrangement for ball bearing and cylindrical roller bearing

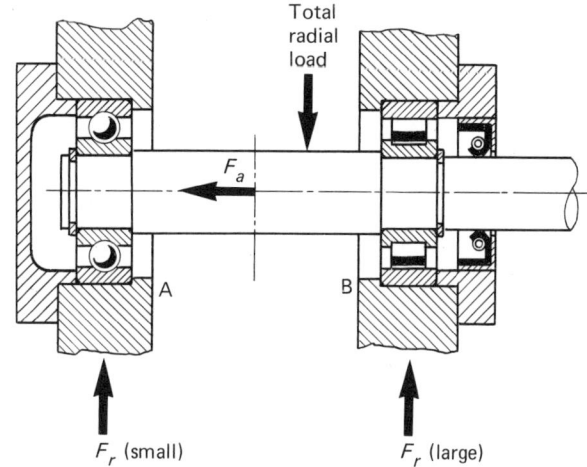

provided. This allows the shaft to expand and contract with changes of temperature.

2 *One Ball Bearing and One Cylindrical Roller* Cylindrical roller bearings are designed to take heavy radial loads but not axial. The arrangement shown in Fig. 5.15 is useful, where the radial load is heavier at one end than at the other.

Ball bearing A locates the shaft and takes all the axial loading. Cylindrical roller bearing B is located closest to the radial load and thus takes more of this radial load than the ball bearing. No thrust is taken by the cylindrical roller bearing, whose design allows for axial expansion and contraction of the shaft.

Fig 5.16 Two taper
roller bearings

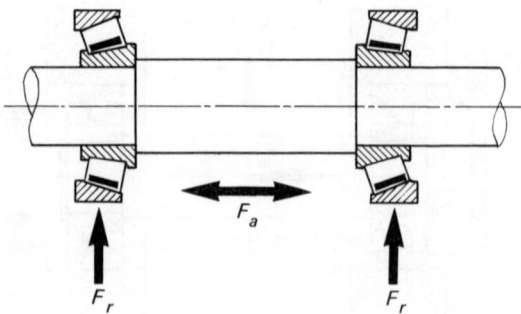

3 *Two Taper Roller Bearings* Taper roller bearings are designed to take heavy axial loads in one direction only. Thus the bearings are arranged with opposed tapers, as shown in Fig. 5.16. Accurate location of each bearing is essential. Their shape imposes an additional axial thrust onto themselves which may assist or oppose the applied axial load F_a. Selection of taper roller bearings is thus a very complex matter, and instructions in manufacturer's catalogues should be strictly adhered to. A similar arrangement for medium duty loads may be used with angular contact bearings.

Fig 5.17 Two
self-aligning bearings

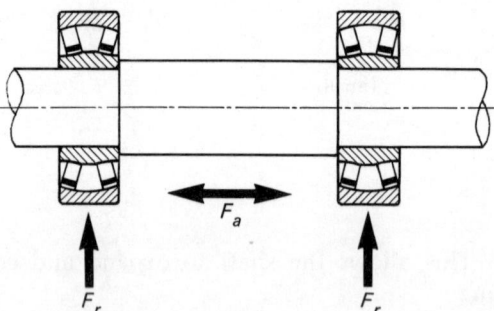

4 *Two Self-aligning Bearings* Self-aligning bearings permit small angular misalignment between shaft and housing due to the spherical grooves in their races. They are particularly useful in taking account of errors in mounting and deflection of slender shafts. Fig. 5.17 shows two spherical roller bearings for heavy radial and thrust loads. A similar arrangement for medium duty loads may be used with self-aligning ball bearings.

5.6
Gear Selection

BS436 Parts I and II lists standard involute gear proportions and these were fully covered in *Engineering Design for Technicians*. A brief revision of terms follows.

Involute = Basic curve for most modern gear forms
Pitch circle diameter (PCD) = True meshing diameter
PCD of driving pinion $= d$
PCD of driven gearwheel $= D$

Centre distance $= \frac{1}{2}(D + d)$

Module $M = \dfrac{\text{PCD}}{\text{No. of teeth } N}$

$$= \frac{D}{N_g} = \frac{d}{N_p}$$

where N_g = number of teeth in gearwheel
N_p = number of teeth in pinion

Addendum = Module = Distance from outside diameter to PCD

Dedendum = $1.25 \times$ Addendum

= Distance from PCD to root diameter

Full depth of tooth = Addendum + Dedendum

= Outside radius − Root circle radius

Circular pitch $C = \dfrac{\pi D}{N} = \pi M$

Circular tooth thickness = $C/2$
Pitch point = Point of pure rolling action on PCD
Base circle = Circle on which involute curve is constructed
Line of action = Direction of resultant tooth load tangential to the base circle and through the pitch point
Pressure angle ψ = Angle between line of action and common tangent to pitch circles (ψ usually equals 20°)

Gear Design

It is not possible to cover design procedures for all common types of gearing in this book. The aim of the following sections is to give an introduction to this complex subject by investigating the "parent" involute gear form— namely the spur gear. Types of gear are illustrated on page 96 and are treated in *Engineering Design for Technicians*.

DRIVING
PINION
DRIVEN
GEARWHEEL

INVOLUTE
INVOLUTE

DEDENDUM
ADDENDUM
P.A.
(ψ)
LINE OF
ACTION
INVOLUTE CURVE
COMMON
TANGENT
OUTSIDE
CIRCLE
PITCH
CIRCLE
ROOT
CIRCLE
A
B
C
D
E
θ
PITCH POINT
CLEARANCE
BASE
CIRCLE
CIRCULAR
TOOTH
THICKNESS
CENTRE DISTANCE

GEAR FORMS
SPUR
HELICAL
DOUBLE
HELICAL
BEVEL
SPIRAL
BEVEL
WORM &
WHEEL
RACK &
PINION

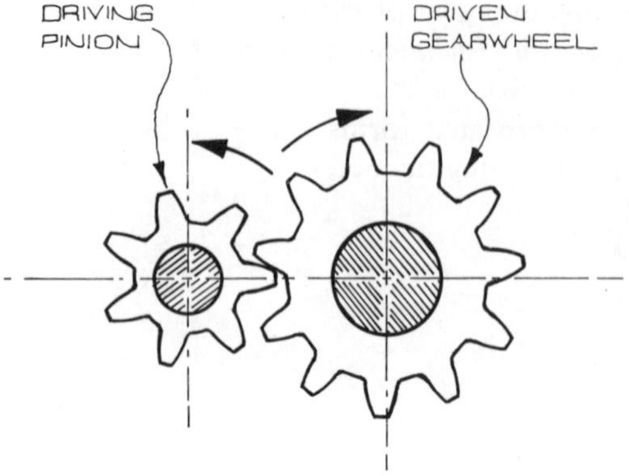

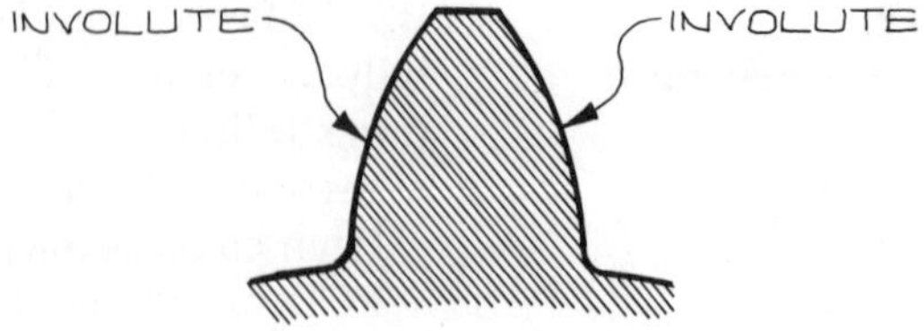

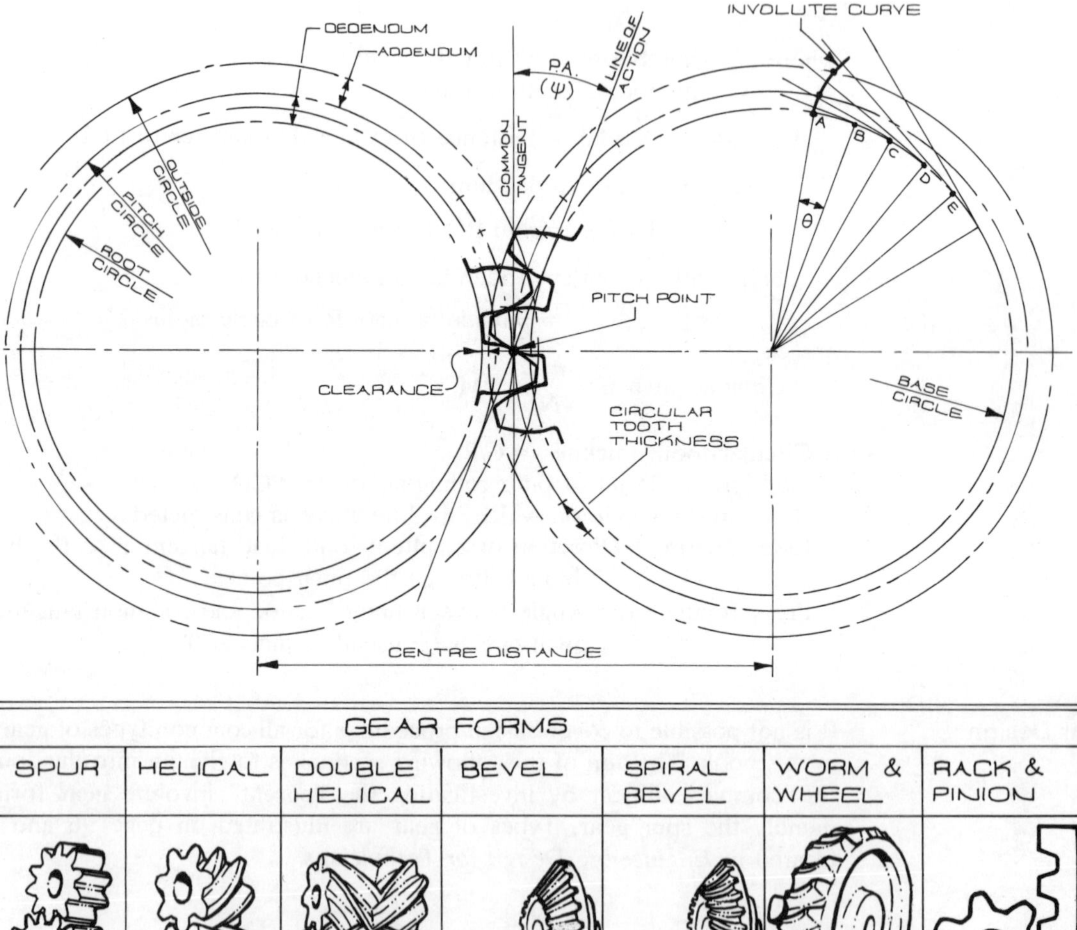

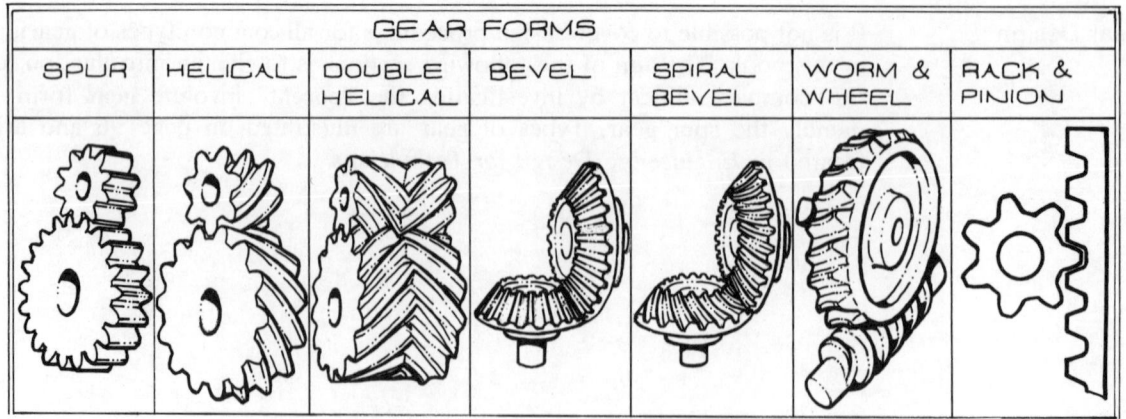

Fig 5.18 Tooth breakage load

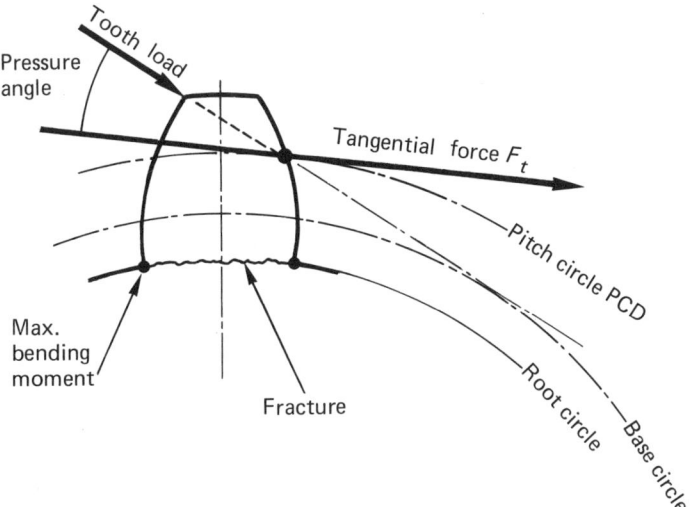

Types of Gear Failure

Gears generally fail through
1) Tooth breakage
2) Tooth wear.

1 *Tooth breakage* occurs near the root of the tooth due to a bending moment applied at the tip of the tooth (Fig. 5.18). The bending moment is caused by the tooth load which acts along the line of action, tangentially to the base circle. Each tooth thus becomes a small cantilever beam with a maximum bending moment near the root circle.

Design calculations to avoid tooth breakage are often called Beam Strength calculations, since it is the tensile strength or fatigue limit of the material which is of paramount importance.

2 *Tooth wear* occurs when pitting and scoring of the tooth surface blemishes eventually form cracks which spread and finally cause small pieces to break away from the parent metal. Causes of tooth wear may be numerous and complex, but obvious considerations include hardness of material surface, quality of surface finish, and lubrication requirements.

Gear-loading Capacity

Numerous formulae and standard procedures have been developed for the calculation of required gear sizes for given loading conditions. The relevant British Standard (BS436, Part 3) was written in 1940 and is now considered to be obsolete. Others commonly used are the American A.G.M.A. and the German D.I.N. standards. B.S.I. now plan to issue a new standard based on recommendations of the I.S.O. (International Standards Organisation).

Meantime the available standards are somewhat inconsistent and we will thus return to basics by quoting a traditional law, namely the Lewis Formula, for beam strength. This is now too primitive for consideration in modern gear philosophy, but will suffice as a simple introduction to the subject.

In the examples considered here it will thus be assumed that teeth have adequately hardened to account for wear and that beam strength is the critical consideration.

Gear Beam Strength Calculations

The Lewis Formula is

$$F_t = YbC\sigma$$

where F_t = tangential tooth load (N)
b = width of teeth (mm)
C = circular pitch (mm)
σ = stress in material (N/mm²)
Y = form factor depending on shape and number of teeth.

The Y factor for standard involute teeth with a 20° pressure angle may be taken as approximately

$$Y = 0.154 - \frac{0.912}{N}$$

where N is the number of teeth.

1 The *allowable stress* σ will be obtained from the tensile or compressive strength of the material (whichever is the smaller) and may be given as:

$$\text{Allowable stress } \sigma = \frac{\text{Tensile or compressive strength}}{\text{SF} \times \text{SP} \times K_t}$$

where SF = safety factor for general conditions of operation
SP = speed factor accounting for the increase in working stresses which occur at higher speeds
K_t = stress concentration factor.

2 The *speed factor* (SP) varies with different types of gear and speed ranges, but, for general applications in metallic spur gears, may be taken as

$$\text{SP} = \frac{3000 + V}{3300}$$

where V = circumferential velocity at pitch circle (mm/s).

3 The *Stress concentration factor* K_t accounts for the fatigue action at the root of the tooth (Fig. 5.19). This will inevitably get more severe as the tooth fillet radius r_0 is reduced (see p. 87).

Table 5.2 gives some approximate K_t values for ratios of root fillet radius against circular pitch (r_0/C).

Table 5.2

r_0/C	0.04	0.06	0.08	0.1	0.13	0.16	0.19	0.22
K_t	2.5	2.2	1.9	1.7	1.6	1.5	1.45	1.4

Fig 5.19

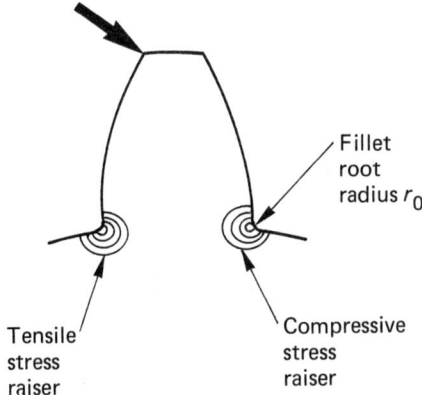

Fillet root radius r_0

Tensile stress raiser

Compressive stress raiser

As a guide, two commonly used r_0/C ratios are 0.0938 and 0.124 (K_t values interpolated from table).

4 The *width of teeth b* is usually fixed in terms of the circular pitch. For general applications a value of 3 to 4 times the circular pitch may be used. This would usually decrease for low-speed gears and increase for high-speed.

Minimum Number of Gear Teeth

If the number of teeth in the pinion or gear is below a minimum, the good contact and freedom from undercutting cannot be maintained. For standard involute teeth with a 20° pressure angle, this minimum may be taken as about 17 teeth.

Example An application of the Lewis gear strength formula is given in Chapter 10 (p. 195).

5.7 Fabricated Steel Structural Sections

These are rolled-steel sections provided to standard sizes and shapes. The calculation of the I values is sometimes complex and these are thus usually obtained from charts supplied for the standard section. Typical sections are shown in Table 5.3. Also, standard sizes available from the British Steel Corporation (B.S.C.) are supplied in charts 4 and 5 at the back of the book.

Rolled flanged sections such as channel and rolled steel joists (RSJ) are listed in BS 4 Part 1 and are widely used in medium/heavy structural and mechanical applications. They are most economical in cases of pure bending in one plane. Hollow steel sections are listed in BS 4848 Part 2, and are useful in light/medium structural and mechanical applications. Because of their high polar second moment of area, they are often used in cases of torsional stress and combined bending and torsion. They are generally lighter and more rigid than equivalent flanged sections. Also their smaller external surface area creates an economic advantage in requiring less painting.

Table 5.3

Dimensions in mm	60×60 (X–X)	80×40	90×90 x5 THK.	120×60 x5 THK.	120×60 x5 THK.	φ70	φ89 x5 THK.	φ114 x5 THK.	102×44
BENDING Relative stiffness	36	44	45	51	33	33	26	45	30
BENDING Mass/metre (kg)	28	25	13.3	13.3	13.3	30	10	13.5	7.44
BENDING Ratio	1.3	1.8	3.4	3.8	2.5	1.1	2.6	3.3	4
TORSION Relative stiffness	51	47	65	56	56	67	52	90	28
TORSION Mass/metre (kg)	28	26	13.3	13.3	13.3	30	10	13.5	7.5
TORSION Ratio	1.8	1.8	4.9	4.2	4.2	2.2	5.2	6.7	3.7

The chief disadvantage of hollow sections is the relative inaccessibility created. Internal threaded fastenings and welded fixtures are not possible without making access holes (and thus reducing the strength). They are also prone to corrosion on internal surfaces, which are difficult to paint. Further comparisons of structural sections are made on pages 102–105.

Structural Steels

These are specialised grades of steel which are listed in BS 4360. In the example in this chapter we will assume the use of Grade 43/A, which has similar strength properties to a mild steel such as 070M20 (EN3), i.e. tensile and compressive strengths 460 N/mm^2 in the example considered.

Example 5.6 *Selecting a Structural Cross-section*
Select a suitable size of rectangular steel tube to support the bending moment shown in Fig. 5.20. Required safety factor is 5.

Fig 5.20

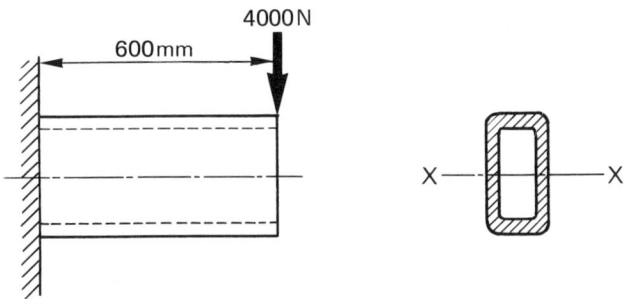

Tensile/compressive strength $= 460 \text{ N/mm}^2$ (see previous note)

Allowable stress $\sigma = 460/5 = 92 \text{ N/mm}^2$

Maximum bending moment $M = 4000 \times 600 = 2.4 \times 10^6 \text{ N mm}$

Since $M/I = \sigma/y$

$$\frac{I}{y} = \frac{M}{\sigma} = \frac{2.4 \times 10^6}{92} = 26087 \text{ mm}^3 = 26.087 \text{ cm}^3$$

Now, the I/y value is also called the Elastic Modulus. Referring to the B.S.C. catalogue extract in chart 5, it is seen that a rectangular tube, size $90 \text{ mm} \times 50 \text{ mm} \times 5 \text{ mm}$ thick, appears to be a suitable cross-section, with an elastic modulus of 28.9 cm about axis X–X. (Note the distinct advantage in using an increased depth of section in the plane of bending; see also pp. 100 and 102.)

Deflection and Stiffness

In many applications, such as machine tool transmissions and large structures, deflection considerations may be just as important as the maximum stress induced. Serious misalignments and interferences caused by excessive deflection could cause a machine to malfunction long before it fractured due to stress. Deflection values are also a useful tool in analysing average strength in structures since the two properties are inversely proportional.

The stiffness value of a design takes account of the loading exerted, and is given as

$$\text{Stiffness} = \frac{\text{Force}}{\text{Deflection}}$$

For example, a stiff structure would have a small deflection for a certain force applied.

It is useful to note that stiffness is directly proportional to strength and thus may be used to compare average stress values of designs.

Strength/Cost Optimization (Strength/Mass Ratios)

Any design may, of course, be made adequately strong and stiff by using excessive amounts of material but this inevitably results in costly and cumbersome structures. Strength and stiffness values should thus be compared with relative mass values in order to achieve the optimum design for materials cost.

This gives a strength comparison ratio of

$$\text{Strength/mass ratio} = \frac{\text{Strength}}{\text{Mass}}$$

Similar comparisons are achieved by considering stiffness:

$$\text{Stiffness/mass ratio} = \frac{\text{Stiffness}}{\text{Mass}}$$

Thus, for example, a high stiffness/mass ratio would indicate an economical design form where deflection is a critical consideration. Table 5.3 compares the stiffness/mass ratios for various shapes of steel sections in bending and in torsion. Investigation of this table reveals that solid sections are the most uneconomical, both for bending and torsion.

Circular shapes give the best results in torsion with tubes vastly superior to solid bar and increasingly so with larger diameters. In the case of shafts transmitting torque however, production and space requirements often limit the use of circular tube. Sections of increased depth are useful in bending but are unsuitable for torsion.

Square tube is generally superior to circular tube in bending and gives adequate results in torsion.

Flanged sections such as the RSJ shown give the most favourable results in bending, but are inferior to tube in torsion.

Box Sections

Box sections have very high stiffness/mass ratios in torsion and also are acceptable in bending. They are thus widely used in cast and fabricated components such as transmission bedplates, machine tool bases and frames, etc., where torsional stiffness is an important consideration.

Fig. 5.21 compares stiffness/mass ratios of thin plate, thick plate and box section for torsional loading. The box section is shown to have the highest ratio and is thus the most economical form for materials cost.

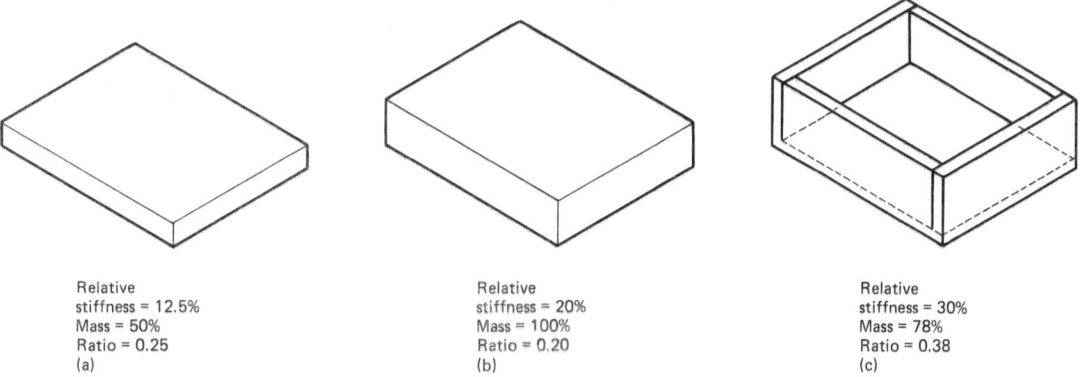

Relative
stiffness = 12.5%
Mass = 50%
Ratio = 0.25
(a)

Relative
stiffness = 20%
Mass = 100%
Ratio = 0.20
(b)

Relative
stiffness = 30%
Mass = 78%
Ratio = 0.38
(c)

Fig 5.21

Torsional stiffness/mass ratios may be improved even further by adding diagonal ribs, as shown in Fig. 5.22.

Fig 5.22

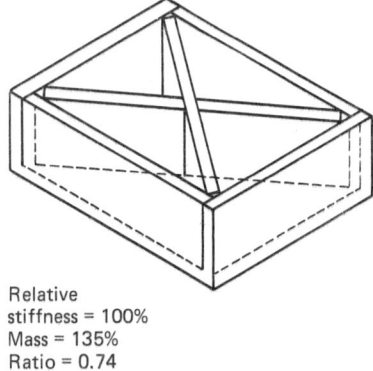

Relative
stiffness = 100%
Mass = 135%
Ratio = 0.74

Fig. 5.23 shows a ribbed box section on the handle of a child's plastic spade. This apparently light-duty item is, in fact, subjected to a great deal of torsional and bending punishment by the enthusiastic youngster. The chosen design produces a light, stiff product with a considerable cost saving over a solid handle, especially in mass production.

Fig 5.23

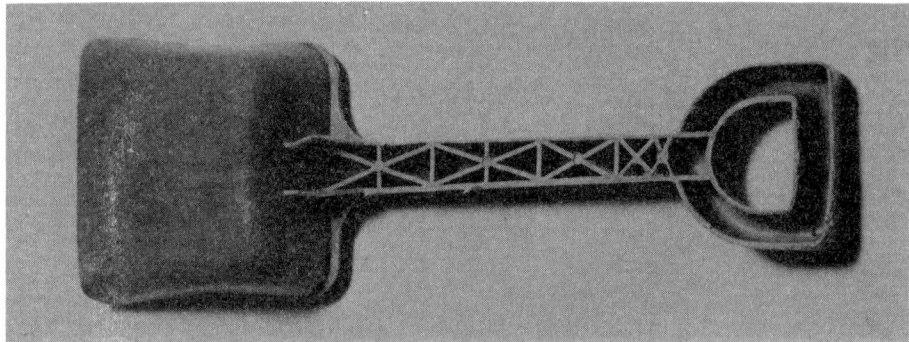

Fig 5.24

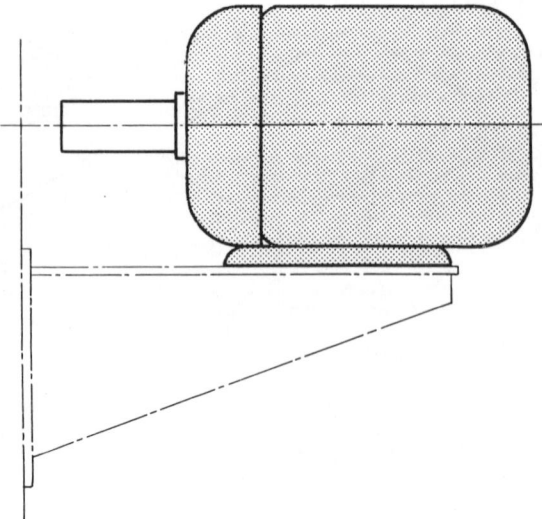

Example 5.7 Design a steel fabricated support bracket to mount the overhung motor shown in Fig. 5.24.

Solution The mass of this component will directly affect its cost. To support an overhung motor, it should possess good stiffness, both in bending and torsion. The alternatives are shown in Fig. 5.25. The stiffness/mass ratios are shown in Table 5.4.

The thick-plate solution (*a*) has the least favourable ratio values and should be rejected. The circular tube solution (*c*) has the most favourable torsional ratio, but is less favourable in bending. In cases of higher torsion, however, this would be the likely choice. The ribbed box section solution (*b*) has the most favourable minimum ratio value and may be chosen on this basis, although the rectangular tube solution (*d*) with average ratios may be preferred for ease of manufacture. (Reliable calculated data may be difficult to achieve here, and the authors strongly recommend the use of scale or full-size test models in the initial comparison stage, discussed in Chapter 7.)

It should be emphasised that materials cost comparisons such as these should be used in conjunction with manufacturing cost considerations and relative costs of forms of supply (e.g. circular tube may be cheaper to buy than rectangular tube, etc.). Also, material wastage should be considered. For example, solution (*c*) requires two holes to be cut in the motor support plate to accommodate the circular tube. This is wasted material and should thus be included in the mass value.

Table 5.4 Stiffness/mass ratios

	(a)	(b)	(c)	(d)
Mass %	100	74	78	72
Torsional stiffness %	40	85	100	70
Ratio	0.4	1.15	1.28	0.97
Bending stiffness %	60	100	75	90
Ratio	0.6	1.35	0.96	1.25

Fig 5.25

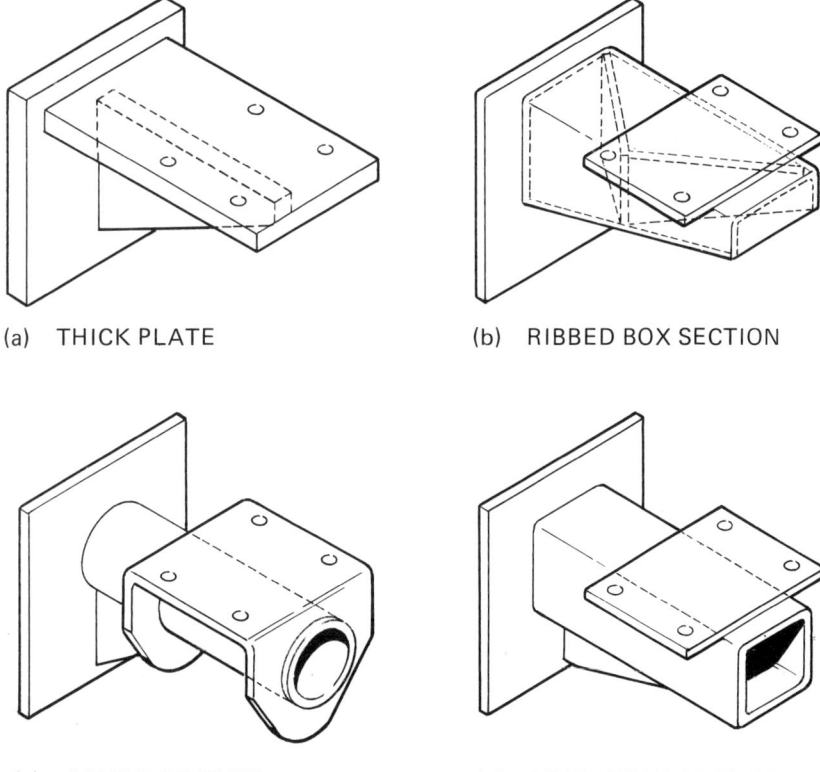

(a) THICK PLATE

(b) RIBBED BOX SECTION

(c) CIRCULAR TUBE

(d) RECTANGULAR TUBE

**5.8
Struts**

Fig 5.26

A strut is a long, slender item which is subjected to compressive loading and is thus likely to buckle at the centre (Fig. 5.26). The shape of a strut may be described by its *slenderness ratio*, which is really the ratio between the overall length and the least radius of gyration of section:

$$\text{Slenderness ratio} = \frac{L}{k}$$

where L = total length.
k = least radius of gyration.

For shorter struts (slenderness ratio up to about 80 for steel), the most commonly used law is the Rankine–Gordon formula, which states:

$$F_r = \frac{A\sigma_c}{1 + a(L/k)^2}$$

where F_r = crushing load (N)
A = area of section (mm^2)
L = total length (mm)
k = least radius of gyration (mm)
a = constant
σ_c = compressive stress at yield of material (N/mm^2)

For structural steel the σ_c value may be taken as approximately 320 N/mm^2. The value of constant a will depend on the type of material and the design of the ends. Table 5.5 gives a values for structural steel.

For slenderness ratios above 80, the Euler formula is often used:

$$F_r = \frac{a\pi^2 EI}{L^2}$$

where a is dependent only on the design of the ends (Table 5.5)
E = Young's Modulus for the material
I = least value of 2nd. moment of area of the cross-section (mm^4).

For structural steels, E may be assumed to be approximately 200 000 N/mm^2.

Circular steel tube is commonly used in strut applications because of its uniformity of bending strength in all planes.

Examples 5.8 The support link in Fig. 5.27 is subjected to a compressive load of 110 000 N. Choose a suitable section of steel circular tube if the required safety factor is 5.

Solution Assuming the Rankine–Gordon formula
for two hinged joints, and structural steel,

$$a = \frac{1}{7500} \quad \text{and} \quad \sigma_c = 320 \text{ N/mm}^2 \quad \text{(see Table 5.5)}$$

Table 5.5 Constant *a* values for structural steel on Rankine–Gordon and Euler strut formulae

	CASE 1	CASE 2	CASE 3	CASE 4
Rankine–Gordon *a*	1/7500	1/15 000	1/30 000	1/1875
Euler *a*	1	2	4	0.25
σ_c (N/mm²)	320	320	320	320
E (N/mm²)	200 000	200 000	200 000	200 000

CASE 1 Both ends hinged.
CASE 2 One end hinged, one end fixed.
CASE 3 Both ends fixed.
CASE 4 One end fixed, one end free.

Fig 5.27

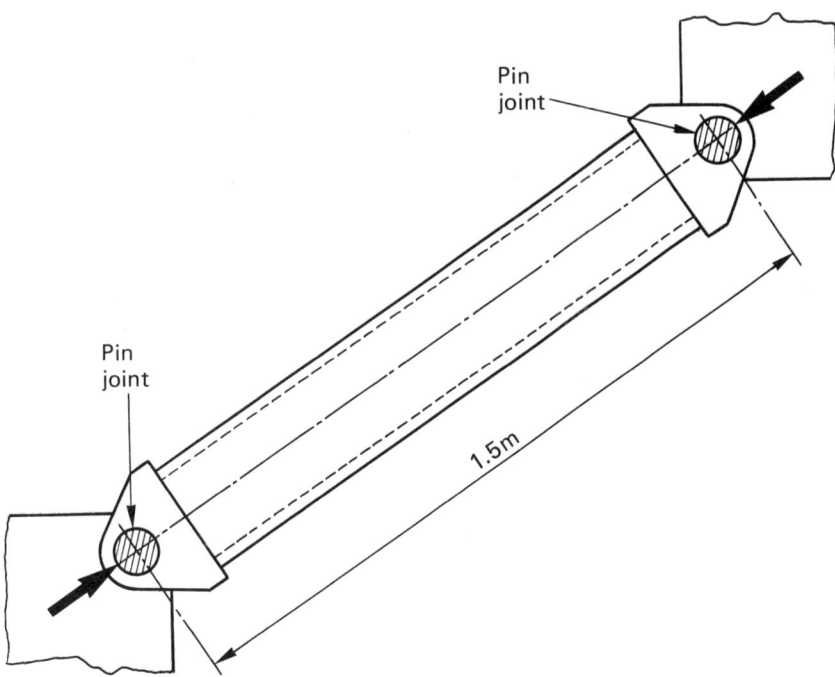

Allowable σ_c (accounting for safety factor) is

$$320/5 = 64 \text{ N/mm}^2$$

Now $\quad F_r = \dfrac{A\sigma_c}{1 + a(L/k)^2}\quad$ which may be transposed to

$$k = \sqrt{\left[\frac{aF_r L^2}{A\sigma_c - F_r}\right]} \quad .$$

To obtain a meaningful answer here, it is necessary that $A\sigma_c$ be greater than F_r or, in other words, that $A > F_r/\sigma_c$. A good selection starting point may therefore be obtained by calculating the F_r/σ_c value and initially choosing the smallest A value which is greater than this. Now,

$$F_r/\sigma_c = 110\,000/64 = 1719 \text{ mm}^2 = 17.19 \text{ cm}^2$$

Referring to B.S.C. Circular Hollow Sections, chart 5 (p. 222), the lowest A value above 17.19 cm^2 is 17.2 cm^2, this being for size 114.3 mm o/d $\times$ 5 mm thick. Now, for $A = 1720 \text{ mm}^2$,

$$k = \sqrt{\left[\frac{(1/7500) \times 110\,000 \times 1500^2}{(1720 \times 64) - 110\,000}\right]} = 642.3 \text{ mm}$$

Thus the minimum allowable radius of gyration for this section is 64.2 cm. When this is compared with the actual radius of gyration of 3.87 cm, it is seen as inadequate.

Now, the next A value up is seen as 21.2 cm^2, this being for size 139.7 mm o/d $\times$ 5 mm thick. For $A = 2120 \text{ mm}^2$,

$$k = \sqrt{\left[\frac{(1/7500) \times 110\,000 \times 1500^2}{(2120 \times 64) - 110\,000}\right]}$$

$$= 35.84 \text{ mm} = 3.584 \text{ cm (minimum allowable)}$$

The actual radius of gyration for this section is shown as 4.77 cm or 47.7 mm.

Checking for Rankine–Gordon formula suitability:

$$\text{Slenderness ratio } L/R = \frac{1500}{47.7} = 31.4$$

This falls within Rankine–Gordon formula range.

The chosen section is 139.7 mm outside diameter $\times$ 5 mm thickness, circular hollow section (to BS 4848 Part: 2).

Example 5.9 A tall crane gantry is supported by four columns, each of which sustains a compressive load of 190 kN, with a required safety factor of 6. If the height of each column is 12 m with both ends assumed to be fixed, choose a suitable size of steel channel or RSJ for the columns.

Solution Assuming the Euler formula:

$$a = 4 \qquad E = 200\,000 \text{ N/mm}^2 \quad \text{(see Table 5.5)}$$

In this formula, the safety factor is applied to the load value:

$$\text{Design load } F_r = 190\,000 \times 6 = 1.14 \times 10^6 \text{ N}$$

Now $F_r = a\pi^2 EI/L^2$ and therefore

$$I = \frac{F_r L^2}{a\pi^2 E} = \frac{1.14 \times 10^6 \times 12\,000^2}{4 \times \pi^2 \times 200\,000}$$

Minimum allowable I value is

$$20.79 \times 10^6 \text{ mm}^4 = 2079 \text{ cm}^4$$

Referring to B.S.C. steel sections on chart 4 (p. 220) it is seen that the highest figure for the "least-value-of-I" for channels is 628.6 cm^4, which is inadequate.

Referring to the table of RSJs (chart 4) it is seen that size 254 mm $\times$ 203 mm has a least-value-of-I of 2278 cm^4 and would thus be adequate.

Problems

The following questions require reference to materials charts 1, 2 and 3 and standard steel section charts 4 and 5 at the back of the book.

5.1 Determine suitable shaft diameters for the following cases:

a) Material: 080M40 (EN8) steel
Torsional loading: 60 000 N mm torque
Safety factor: 7
Shaft has a standard keyway.

b) Material: 070M20 (EN3) steel
Bending load: 3.4×10^6 N mm maximum bending moment
Safety factor: 8

c) Material: 817M40 (EN24) alloy steel
Combined bending and torsion load:
1.6×10^6 N mm torque, 5.5×10^6 N mm maximum bending moment
Safety factor: 6
Shaft has a standard keyway.

[*Ans. a*) 19.75 mm (say 20 mm), *b*) 84.45 mm (say 85 mm), *c*) 62.9 mm (say 65 mm)]

5.2 A shouldered shaft made of 080M40 (EN8) steel is required to sustain a maximum bending moment of 10 000 N m and torque of 7000 N m under cyclic loading. The shoulder is to be used as a bearing location with a required safety factor of 2.5 on the fatigue limit.

a) Choose suitable proportions of shaft diameter, shoulder diameter and fillet radius by reference to Table 5.1 and to bearing catalogues.

b) Choose a suitable fatigue stress concentration factor K_t from Table 5.1.

c) Determine suitable dimensions for the shaft.

5.3 A keyed shaft made from 817M40 (EN24) steel is subjected to a cyclic torque loading of 7000 N m with a required safety factor of 4 on fatigue limit. If the keyway is end-milled, and close to a shoulder, choose suitable proportions and stress concentration factor from Table 5.1 and thus determine suitable dimensions for this area of the shaft.

5.4 The following figures were obtained for a spherical ball bearing from a catalogue:

Dynamic load rating $C = 80\,000$ N
Radial X factor $= 0.5$
Axial Y factor $= 1.3$

Determine the expected L10 hours life of the bearing for an applied radial load of 20 000 N, axial thrust load of 8000 N and speed of 160 rev/min.

[*Ans.* 6282 hours (L10)]

5.5 A 20-tooth spur gear of module 6 mm, root fillet radius 2.5 mm, and pressure angle 20°, is required to transmit 30 kW of power at 500 rev/min. The chosen material is 080M40 (EN8) steel and required safety factor is 2.5. Determine the minimum allowable width of teeth on the basis of beam strength. (See Chapter 10, p. 195.)

[*Ans.* 56.14 mm]

5.6 The spur gear in **5.5** is fixed mid-position between two bearings 160 mm apart at centres, on a keyed shaft made of 080M40 (EN8) steel.

a) In addition to the calculations made in **5.5** determine also the tooth separating force, the resultant tooth force, the bearing reactions, and the maximum bending moment.

b) Hence determine the minimum allowable diameter of shaft for a required safety factor of 8 (non-cyclic loading). (See Chapter 10, p. 194.)

[*Ans.* 42.92 mm]

5.7 Choose a suitable size of standard rectangular steel tube which will sustain a maximum bending moment of 7.5×10^6 N mm with a required safety factor of 8.

[*Ans.* 160 mm $\times$ 80 mm $\times$ 8 mm (bending about neutral axis XX)]

5.8 A length of circular steel tube is to be used as a strut which is hinged both ends. A compressive load of 200 000 N is applied and the required length at hinge centres is 2 m. Choose a suitable size of standard cross-section if the required safety factor is 7.

[*Ans.* 219.1 mm outside diameter $\times$ 8 mm thickness circular tube. (Smallest available cross-sectional area with adequate radius of gyration using Rankine–Gordon Formula)]

6 Anthropometrics and Ergonomics

Anthropometrics may be defined as the listing of data information on human body size.

Engineering Design for Technicians included anthropometry as one of a group of scientific disciplines which came under the general heading of Ergonomics, and thus a brief revision of this subject will be initially undertaken.

Ergonomics is the scientific study of the relationship between people and their working environment. The working environment may be taken to include any factor which could affect the efficiency of the working person. Typical considerations include heat conditions, lighting, fellow colleagues, tools, machines and methods of organisation and production.

6.1
The Man/Machine Relationship

Any person who operates a machine may be considered as part of a closed *control loop* system in which he or she receives and processes information and then acts on it. A typical control loop is shown in Fig. 6.1.

Information (e.g. the speed of a machine) is sent to the operator from a *display element* via the display communication channel. A display is any source of information aiding the operator in the control of the machine. Typical displays include dial gauges, digital read-outs and warning lights.

Fig 6.1 Typical control loop

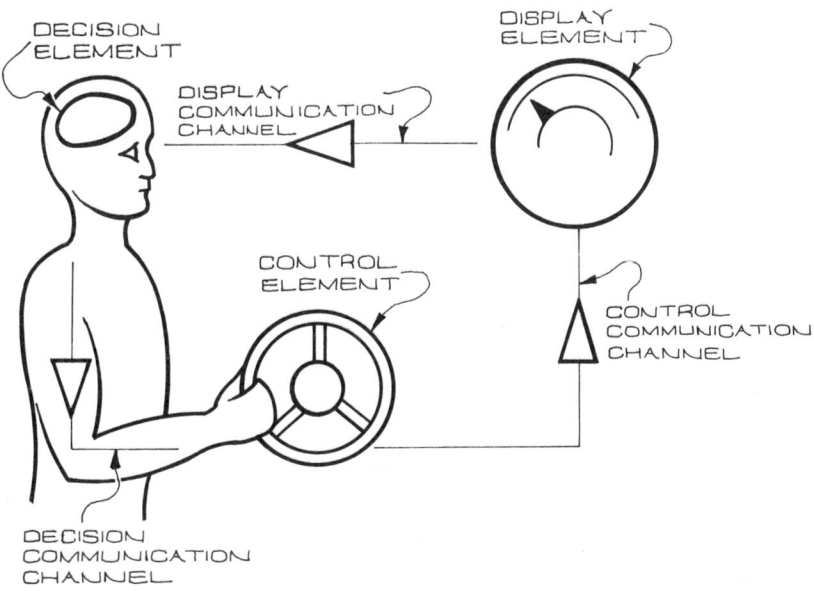

Fig 6.2

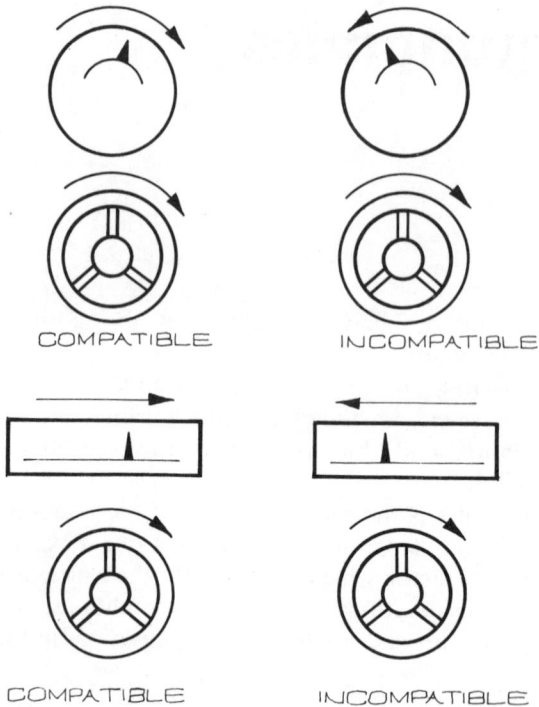

COMPATIBLE   INCOMPATIBLE

COMPATIBLE   INCOMPATIBLE

The information from the display is passed to the control mechanism of the brain via the optical and nervous system where it is processed to arrive at a decision in relation to the required performance. The decision is then communicated to the control element via the mechanical leverage system of the human bone and muscle which makes up the decision communication channel.

A *control* is any device which regulates the action of a machine. Typical controls include handwheels, handles, levers, control knobs, pedals and push buttons. The effect of the action of the machine will be registered on the display via the control communication channel and the loop is thus completed.

The efficiency of the control loop will be affected by various internal and external factors. The display element should be easily and accurately readable and have a movement which is compatible with the movement of the control. Fig. 6.2 shows obvious distinctions between compatibility and incompatibility. The position of displays should be such that they can be communicated to the operator with the minimum of physical effort.

The display communication channel should have a direct line of path to the decision-making organs and be free from interference such as glare on an illuminated panel.

The operator must be physically and mentally capable of making the required decision under satisfactory working conditions.

The decision communication channel should provide easy access to the control and be free from interference. The control element should be of suitable size and easily operated and give compatibility with the display.

The control communication channel must be a reliable mechanism.

External factors such as heating, lighting, noise, ventilation, physical obstructions and fellow colleagues must also have a considerable effect on the control loop.

6.2
The "Average" Person

Physically, people vary in size, weight, strength and shape and in their abilities to see and hear.

As a yardstick, the ergonomic designer must therefore often use average human data. However, anyone with a knowledge of statistics will realise that a simple average or *mean* value can be very misleading and may require further analysis regarding the scatter from this value.

Where mean values only are considered it is likely that only about half of the population under consideration is satisfied by a given design. Thus to ensure that a broad range of population is accommodated, human data which deviates from the mean is often accounted for.

Human data is best analysed by considering a *Normal Distribution or Gaussian Curve*, which plots stated sizes against percentage of population with that size. For example, Fig. 6.3 shows the normal distribution curve for the height of clothed men in the U.K.

Fig 6.3 Normal distribution curve for heights of adult males

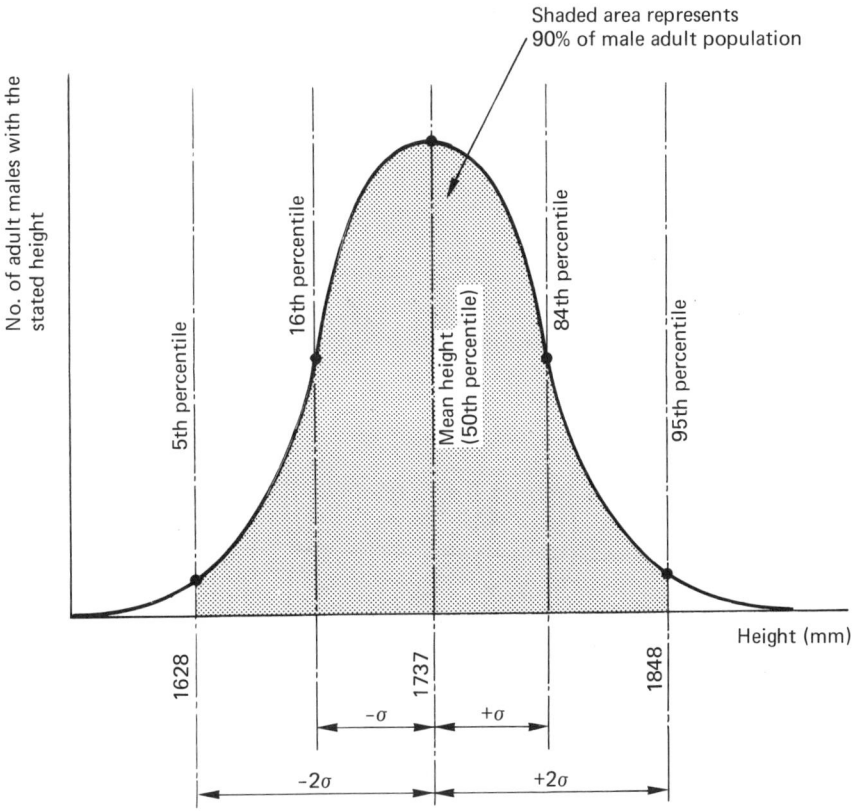

The *standard deviation* σ may be considered as the average amount of scatter above or below the mean height value. It can be assumed to represent about 34% of the population. Thus, the average scatter above or below the mean is equivalent to 2σ or 68% of the population.

Population proportions are often expressed as *percentiles*. For example,

Mean height value $(1737 \text{ mm}) = 50\text{th percentile}$

i.e. 50% of the male population have this height or above.

$$\text{Mean} - \sigma = 50\% - 34\%$$
$$= 16\text{th percentile}$$

i.e. only 16% of the male population have less than this height.

$$\text{Mean} + \sigma = 50\% + 34\%$$
$$= 84\%$$

i.e. 84% of the male population do not exceed this height.

Further analysis reveals that double the average scatter above or below the mean (i.e. mean value $\pm 2\sigma$) gives the 5th and 95th percentiles: i.e. only 5% of the male population do not exceed the 5th percentile height (1628 mm) but 95% of the male population do not exceed the 95th percentile height (1848 mm). The 5th and 95th percentile values thus cover a range equivalent to 90% of the male population.

When considering the full range of population for a design it is thus common practice to take the 5th and 95th percentiles as the lower and upper limits of consideration (see Fig. 6.4). In most cases it would not be practical or economic to consider 100% of the population by catering for

Fig 6.4

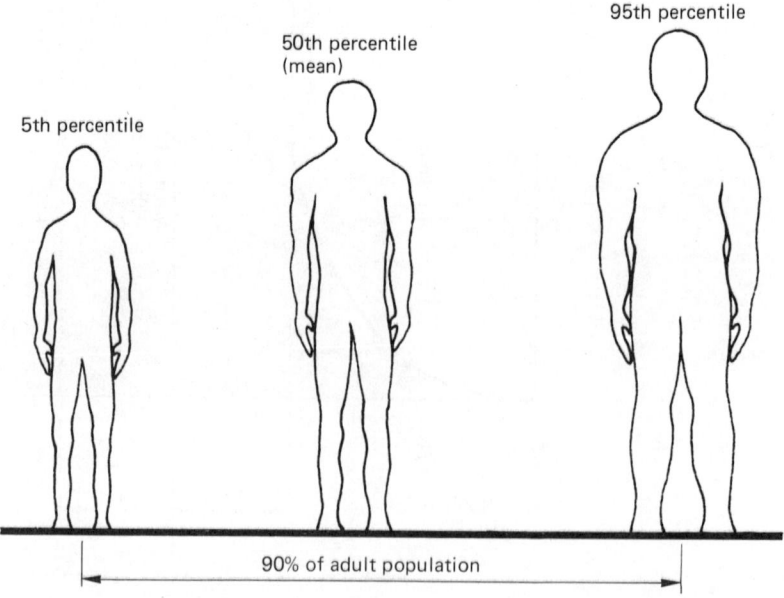

5th percentile

50th percentile (mean)

95th percentile

90% of adult population

people at the extremes and attempts to do so could drastically limit the efficiency of the design for the majority of people.

The 16th and 84th percentiles (mean $\pm\sigma$) are sometimes used if the designer is aiming a product at a smaller or more specialised section of the population.

6.3 Anthropometric Data

In anthropometrics the human data considered is entirely that of body size. Anthropometry thus becomes a scientific discipline in its own right due to the immense variety of dimensions which exist.

Anthropometric data charts may account for total population or, alternatively, may concentrate on specialised groups. Broad group headings may include age, sex, nationality and state of health. The type of data chart used will, of course, depend on the market at which the design is aimed. For example, a certain design of car aimed at the export market could be assumed to be used by adult men and women of a wide range of nationalities. However, a design of children's toy car or disabled person's car would require specialised data.

It should be emphasised that these charts can only give a guide in deciding sizes on a design, and are subject to certain limitations. One of the particular difficulties in complying with anthropometric data is in obtaining reliable values of arc movements for the human frame. It is thus strongly recommended that the charts be used in conjunction with pin-jointed scale models of the mean, lower percentile and higher percentile values of human frame size. Three-dimensional models would, of course, be ideal but may be costly. In many cases two-dimensional cardboard or plastic models of front and side elevations would suffice (Fig. 6.5).

Even when such models are used, complete satisfaction for the required range of users cannot always be guaranteed since the pin joints shown are arbitrarily decided as being the centres of arc movement. In fact human body movements are extremely complex and may act not just from a few

Fig 6.5 Two-dimensional cardboard manikin models

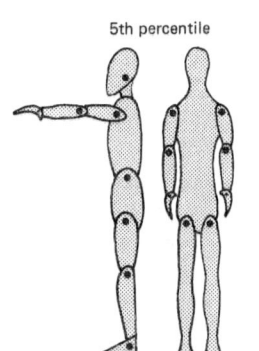

5th percentile

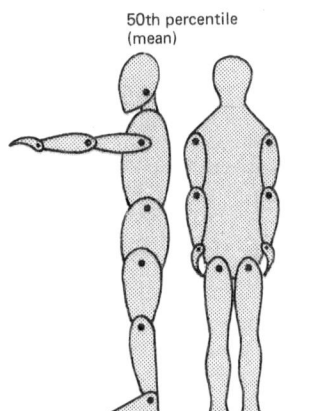

50th percentile (mean)

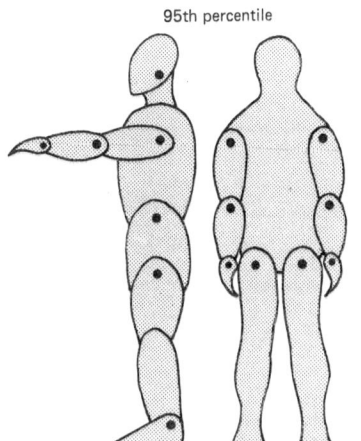

95th percentile

isolated fulcrum points, but as the resultant effect of many movements which affect the shape of the whole body structure.

In the final analysis it may be necessary to test a prototype, or full-scale mock-up, of the proposed design with live subjects. These subjects will ideally conform to the mean, lower percentile and higher percentile of the relevant anthropometric data for the design.

Table 6.1 Anthropometric data chart (supplied by *Architect's Journal*)

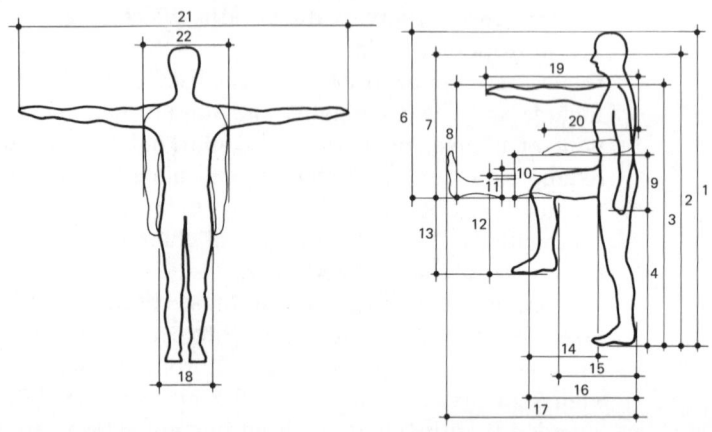

Key dimensions	Men aged 18 to 40 Percentiles			Women aged 18 to 40 Percentiles			Elderly women aged 60 to 90 Percentiles			Examples of applications to design problems
	5th	50th	95th	5th	50th	95th	5th	50th	95th	
Standing										
1 Stature	1628	1737	1846	1538	1647	1758	1454	1558	1662	95th: Minimum floor to roof clearance: allow for headgear, say 100 mm in appropriate situations.
2 Eye height	1524	1633	1742	1437	1546	1655	1338	1451	1564	50th: Height of visual devices, transoms, notices etc.
3 Shoulder height	1328	1428	1428	1237	1333	1429	1201	1288	1375	5th: Height for maximum forward reach.
4 Hand (knuckle) height	703	770	837	—	—	—	653	732	811	95th: Maximum height of grasp points for lifting.
5 Reach upwards	1972	2108	2284	—	—	—	1710	1852	1994	5th: Maximum height of controls: subtract 40 mm to allow for full grasp.
Sitting										
6 Height above seat level	841	900	959	790	849	908	739	798	857	95th: Minimum seat to roof clearance; allow for headgear (men 75 mm, women 100 mm) in appropriate situations.
7 Eye height above seat level	726	785	844	676	735	794	621	684	747	50th: Height of visual devices above seat level.

Table 6.1 Continued

Key dimensions	Men aged 18 to 40 Percentiles			Women aged 18 to 40 Percentiles			Elderly women aged 60 to 90 Percentiles			Examples of applications to design problems
	5th	50th	95th	5th	50th	95th	5th	50th	95th	
8 Shoulder height above seat level	537	587	637	494	544	594	479	529	579	5th: Height above seat level for maximum forward reach.
9 Lumber height	—	254	—	—	—	—	—	—	—	50th: Height of table above seat.
10 Elbow above seat level	178	224	270	157	203	249	143	193	243	50th: Height above seat or armrests or desk tops.
11 Thigh clearance	124	149	174	121	146	171	93	131	169	95th: Space under tables.
12 Top of knees, height above floor	506	552	598	473	519	565	460	498	536	95th: Clearance under tables above floor or footrest.
13 Underside thigh, height above floor	402	435	468	385	418	551	366	404	442	50th: Height of seat above floor or footrest.
14 Front of abdomen to front of knees, distance	336	386	436	—	—	—	—	—	—	95th: Minimum forward clearance at thigh level from front of body or from obstruction, e.g. desk top.
15 Rear of buttocks to back of calf, distance	436	478	520	423	465	507	424	470	516	5th: Length of seat surface from backrest to front edge.
16 Rear of buttocks to front of knees, distance	568	614	660	542	584	626	520	579	638	95th: Minimum forward clearance from seat back at height for highest seating posture.
17 Extended leg length	998	1090	1182	—	—	—	892	967	1042	5th (less than): Maximum distance of foot controls, footrest, etc., from seat back.
18 Seat width	328	366	404	353	391	429	321	388	455	95th: Width of seats, minimum distance between armrests.
Sitting and standing										
19 Forward reach	773	848	923	600	675	750	665	736	807	5th: Maximum comfortable forward reach at shoulder level
20 Folded arm Forward reach	483	530	576	375	422	468	415	460	505	50th: Folded-arm reach for hand controls
21 Sideways reach	1634	1768	1902	1509	1643	1777	—	—	—	5th: Limits of lateral finger tip reach; subtract 130 mm to allow for full grasp.
22 Shoulder width	420	462	504	376	418	460	381	431	481	95th: Minimum lateral clearance in work space above waist.

Anthropometric Data Charts— Main Dimensions

Table 6.1, supplied by courtesy of the *Architects' Journal*, gives some dimensions of body and reach for adult clothed men and women in the U.K.

In nearly all situations to which the designer applies anthropometric data, users will be clothed, and therefore the data in this table includes allowances for clothing and shoes. The allowances for footwear are 28 mm for men, 40 mm for women and 31 mm for elderly women. The allowances for clothing, affecting most of the dimensions from item 6 on, range according to circumstance from 3 mm to 20 mm. In situations where clothes are not worn, e.g. bathrooms and shower-rooms, approximate deductions should be made.

6.4 Types of Display

Displays may be broadly listed under the following headings (see Fig. 6.6):

1) Moving pointer, fixed legend
2) Fixed pointer, moving legend
3) Digital counters.

For ease of reading single values, the digital counter is undoubtedly the most efficient. However, for detection of the progress of a particular measured quantity, a pointer-and-dial display would be preferred. A moving pointer is generally more easily read than a fixed pointer, but the latter may prove more efficient over wide ranges of values, where an "open-window" display can give extremely satisfactory results (see Fig. 6.7).

Straight legend displays may be read horizontally or vertically, but most research indicates that more reading errors are caused when the latter is used (see Fig. 6.8).

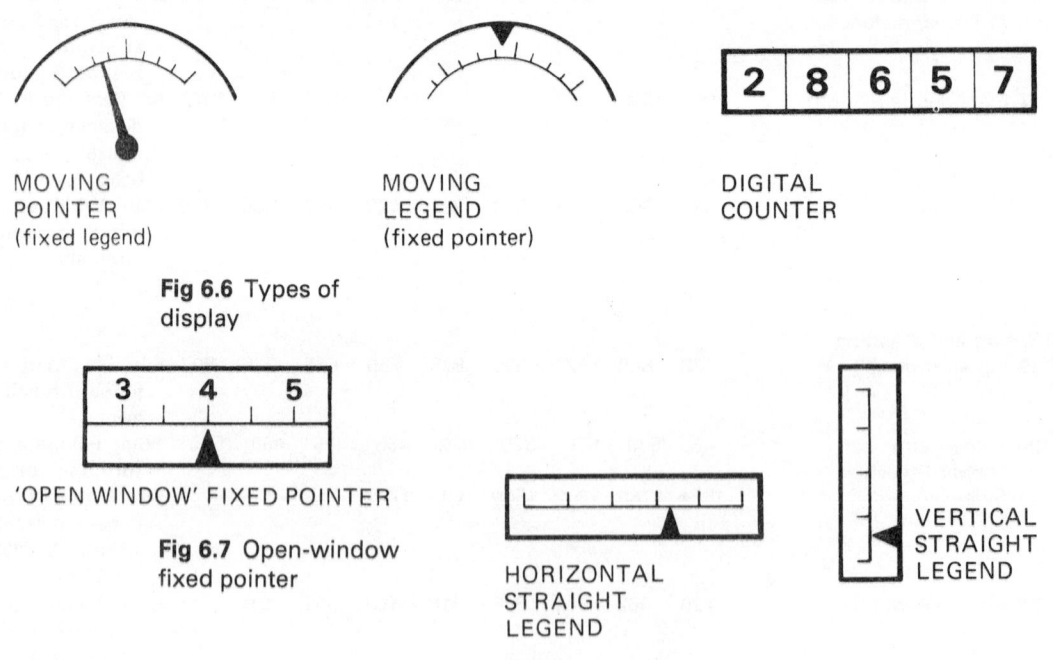

MOVING POINTER (fixed legend)

MOVING LEGEND (fixed pointer)

DIGITAL COUNTER

Fig 6.6 Types of display

'OPEN WINDOW' FIXED POINTER

Fig 6.7 Open-window fixed pointer

HORIZONTAL STRAIGHT LEGEND

VERTICAL STRAIGHT LEGEND

Fig 6.8 Straight legend displays

6.5
Types of Control

(Based on data obtained from J. Croney and K. F. H. Murrell)
Table 6.2 compares the suitability of various controls for different purposes and gives load values which operators would be able to exert. Load capacities in data such as this are usually quoted as the worst possible case, e.g. a standing operator can usually exert greater loads than a sitting operator, and thus the data for the latter case would be quoted.

Hand Controls

The following anthropometric data for the human hand (*courtesy of J. Croney*) may be used as an aid to the design dimensions of hand-controls.

	5th percentile (mm)	50th percentile (mm)	95th percentile (mm)
Hand length (male)	178	190	203
Hand length (female)	165	178	190
Hand breadth (male)	93.5	104	114
Hand breadth (female)	76	89	102

Hand-levers

Levers give a quick control action and can accommodate large forces. They are not suitable for fine adjustments, but can provide efficient ON/OFF or step-by-step control. Handles of levers should give an effective and comfortable grip, and be of sensible size for the human hand. Fig. 6.9 shows the optimum size for human power grip to be in the region of 44 mm diameter.

Fig 6.9 Human hand power grip

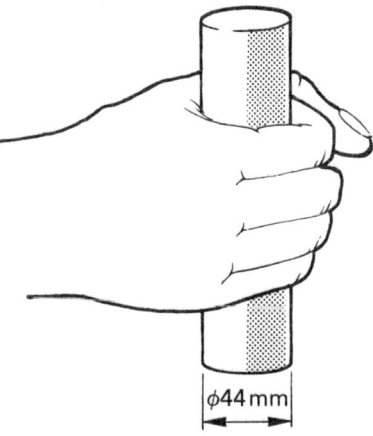

$\phi 44$ mm

Hand-wheels

Hand-wheels provide a controlling torque via both hands of the operator. As with levers, hand-wheels are used when heavy loads are exerted. They can provide good accuracy of adjustment, but are unsuitable where speed is essential. Generally, hand-wheels should have a diameter between 305 mm and 356 mm.

Table 6.2 Controls comparison chart

	Speed	Accuracy	Load Rating	Range	Load Capacity
Lever (horizontal)	Good	Poor	Poor	Poor	67 N upwards 89 N downwards
Lever (vertical)	Good	Medium	Good	Poor	220 N inwards 178 N outwards
Handwheel	Poor	Good	Good	Medium	Up to 17 Nm torque
Small Crank	Good	Poor	Poor	Good	Up to 4.5 Nm torque
Large Crank	Poor	Poor	Good	Good	Above 4.5 Nm torque
Small Knob	Good	Poor	Poor	Medium	
Large Knob	Poor	Good	Poor	Medium	
Step-by-step Switch Knob	Good	Good	Poor	Poor	
Push/Buttons Toggle switch	Good		Poor	Poor	
Pedal (Whole Leg)	Good	Poor	Good	Poor	Up to 900 N
Pedal (Heel Pivot)	Medium	Medium	Poor	Medium	Up to 90 N

Cranks

Cranks are intended to provide torque via one hand. Smaller cranks (up to about 100 mm diameter) are used for fast control. Larger-diameter cranks would give a slower control and increase torque capacity.

Knobs

Rotating knobs are intended for light loading control with either two or three fingers, or with the whole hand. Knobs are typically used in applications such as instrument control panels. The essential requirements of control knobs are that they are of comfortable size, are easily rotated, have an efficient grip, and avoid any information being obscured by the operator's fingers.

Small control knobs (down to about 10 mm diameter) may be used for minimum loadings where speed of adjustment is desired and coarse adjustment is permitted.

Fine sensitive control will require larger diameters (between 35 and 50 mm).

Step-by-step switch knobs require higher turning torques and are thus more efficiently operated with the whole hand. This will require diameters up to 75 mm and sometimes operation by wrist movement.

Fig. 6.10 summarises the types of control knob discussed.

To ensure efficient grip, the control knob should have a depth of around 20 mm. Also, grip is often aided with small serrations.

Data from E. Grandjean indicates that knobs operated by one or two fingers should be not less than 25 mm apart and those operated by the whole hand should be not less than 50 mm apart (based on 95th percentile hands).

COARSE CONTROL
(down to φ10 mm)

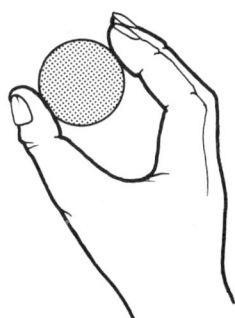

FINE CONTROL
(φ35-φ50 mm)

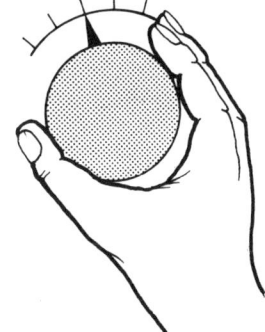

STEP-BY-STEP
CONTROL
(up to φ75 mm)

Fig 6.10 Types of control knob

**Push Buttons and
Toggle Switches
(Fig. 6.11)**

These are essentially used as light-load ON/OFF controls and would normally be designed for operation by one finger, although large-energy push buttons are sometimes designed for use of the whole hand.

One-finger push buttons should be big enough to accommodate a 95th percentile forefinger and thus should not be less than about 12 mm section.

Toggle switches should ideally have a minimum length of 12 mm and an angular movement of 45° (*courtesy of E. Grandjean*).

Fig 6.11

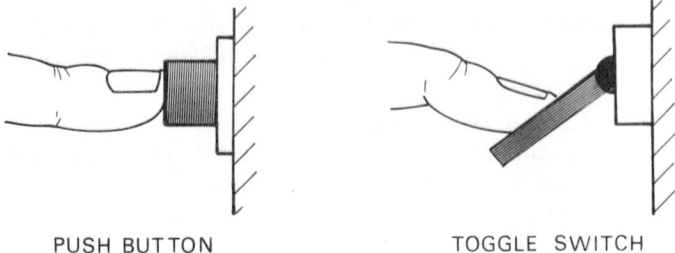

PUSH BUTTON   TOGGLE SWITCH

Fig. 6.12 shows a portable television set produced by Hitachi. The far-right control is a push button for ON/OFF power and the right-of-centre control is a rotating knob for volume control. The left-of-centre control is a rotating knob for fine control of channel tuning. The far-left control is a push button for ON/OFF display of the tuning cursor which moves in conjunction with rotation of the centre tuning knob. Note the difference in size between tuning and volume control knobs, and the distance between the controls.

Fig 6.12

Joysticks

The joystick is a type of hand control now used extensively in computer applications, including computer numerically controlled (CNC) machine tools. It provides a fast, coarse control for light loading in any direction on one surface. Fig. 6.13 shows a close-up of the quill feed control for the Bridgeport CNC milling machine discussed in Chapter 8.

Staying with this milling machine, Fig. 6.14 shows how efficient use of a control panel may be enhanced by colour contrast and function symbols.

Fig 6.13 Joystick control [*Bridgeport Textron*]

Fig 6.14 Control panel with function symbols [*Bridgeport Textron*]

Foot Pedals

These may be used for fast action control with medium/heavy loading capacity. However, they lack the accuracy and range which may be obtained with hand controls.

Pedals may be operated in the seated position, using the whole leg or by pivotting the foot at the heel. They are not intended for use in the standing position.

The following anthropometric data for the human foot (*courtesy of J. Croney*) may be used as an aid to the design dimensions of foot pedals.

	5th percentile (mm)	50th percentile (mm)	95th percentile (mm)
Foot length (male)	249	267	286.5
Foot length (female)	223	241	259
Foot breadth (male)	91	98.5	109.5
Foot breadth (female)	83.5	91	98.5

Pedals operated with the whole leg can produce maximum force for short durations but accuracy is extremely low. High pedal pressures are best achieved with angles of the knee of around 135° and by supplying a back rest.

Pivotting at the heel produces lighter loads but gives more accurate control and is more suitable for longer time durations. Fig. 6.15 shows some suggested dimensions to aid the designer of this type of pedal (50th percentile values) (*supplied by courtesy of K. F. H. Murrell*).

Fig 6.15 Heel-pivot pedal control

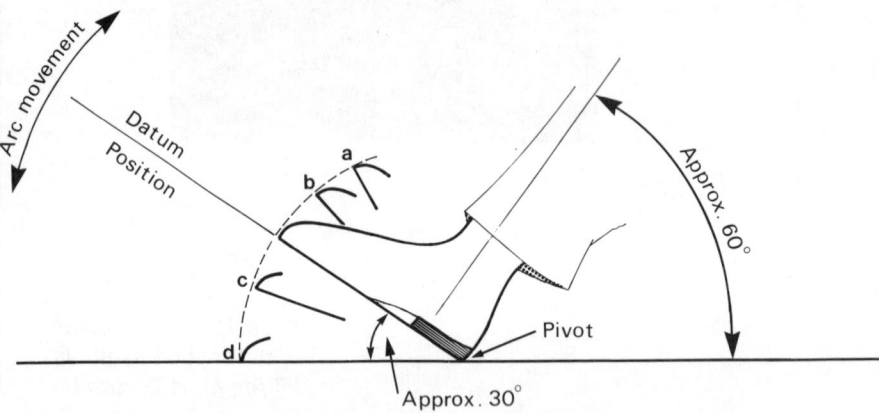

a   Max. arc movement of toe above datum = 114 mm (male), 102 mm (female)

b   Normal arc movement of toe above datum = 76 mm (male), 76 mm (female)

c   Normal arc movement of toe below datum = 114 mm (male), 102 mm (female)

d   Max. arc movement of toe below datum = 152 mm (male), 140 mm (female)

A car interior contains a wealth of anthropometric and ergonomic considerations as indicated in the interior layout of the Ford Sierra 4 X 4, shown in Fig. 6.16. These considerations include:

a) Size, shape, position and material of the steering wheel, gearstick, handbrake, foot pedals and column controls.

b) Shape of facia panel and size, shape and arrangement of all facia controls and displays. Visual clarity of displays and logic of the indication symbols, e.g. petrol gauge, wiper speeds, etc.

c) Allowance for seat adjustment.

d) Compatibility between all controls and displays.

e) Position, size, shape and logic of door handle, window opener, and seat-belt fixing.

(see Problems 6.8 and 6.10 on page 126.)

Fig 6.16 Interior of the Sierra 4 X 4: a good study in anthropometric and ergonomic considerations [*Ford Motor Company*]

Problems

6.1 Distinguish between the terms "anthropometrics" and "ergonomics".

6.2 Explain what is meant by the 5th, 50th and 95th percentile values of anthropometric data and state the significance of these values to design thought.

6.3 State the main limitations of anthropometric data and suggest ways in which these may be overcome.

6.4 State the essential requirements of a good display.

6.5 List some common forms of display and describe their applications, advantages and limitations.

6.6 Design a facia panel of a typical stereo music centre incorporating both display and control. Giving reasons, state the type, size and distance apart for all controls shown. (Pay particular attention to requirements of coarse or fine adjustment.)

6.7 Determine the overall dimensions of a metallurgical microscope for use by seated adults at a desk. Both fast and fine adjustment of focussing and lateral table movement should be supplied and the position and size of controls should be clearly stated. (Assume desk top 720 mm from the ground, ref. BS 5940: Part 1, 1980). If possible a range of 5th percentile women to 95th percentile men should be catered for.

6.8 The foot pedals of a car fall into two distinct categories. Giving logical arguments, state the category to which each pedal belongs. Hence use anthropometric data to design a simple set of car pedals.

6.9 Determine the overall dimensions of the main body of a typewriter for use by male or female operators. Also give recommendations for:

i) The size and distance apart of the keys.

ii) The type, size and dimensional location of the spool-winding control knob.

6.10 For a typical family car decide on:

a) The diameter and dimensional location of the steering wheel with reference to the basic seat position.

b) The position of the dashboard relative to the back of the seat.

c) The arrangement, types, sizes and dimensional locations of an average display of dashboard controls with reference to the centre of the steering column.

e) The type, size and position of window-opening device.

f) The amount of seat adjustment required for a range from 5th percentile women to 95th percentile men.

Refer to Fig. 6.16 on page 125.

Bibliography

E. Grandjean *Fitting the Task to the Man* (Taylor and Francis).

John Croney *Anthropometrics for Designers* (B. T. Batsford).

K. F. H. Murrell *Fitting the Job to the Worker* (British Productivity Seminar on Ergonomics).

The Architects' Journal.

British Standards BS 5940 *Office Furniture* Part 1, 1980.

W. H. Mayall *Industrial Design for Engineers* (Iliffe Books).

7 The Use of Models in the Design Process

7.1

Types of Model

The aircraft designer who makes a simple paper dart model of a new idea to "see if it flies" is demonstrating a classic example of the invaluable contribution that scale models can make to the design process. The designer may have started with some sketches and basic calculations, but the scale model gives an immediate indication of three-dimensional appearance and design performance which could not be matched by drawing or calculation at this initial stage.

Each year in the U.K., millions of pounds are spent in producing scale models of new components. Industrial models fall into numerous categories, but may be broadly listed under four headings:

1) Sales models
2) Layout models
3) Aesthetic/ergonomic models
4) Test models.

1 *Sales Models* can give customers a good indication of the attributes of a new design. Scale models of this type usually take the form of highly accurate "dinky toy" type items made of zinc alloy die castings or plastic mouldings.

2 *Layout Models* are often used to aid communication in the awkward three-dimensional problems which frequently occur in fields such as piping design and plant layout. Scale models made of wood, cardboard or plastic are often superior to drawings in aiding decisions on, for example, the arrangements of machines in a factory, especially if the models are easily re-arranged to consider the alternative ideas.

3 *Aesthetic/Ergonomic Design Models* Industrial modellers are extensively employed by engineering companies to make scale models or full-size models of new designs, in order to develop the visual effect. The importance of appearance on the marketing potential of a new product is outlined in Chapter 8. Many different models may have to be made and modified before the final appearance and shape is chosen.

Fig. 7.1*a,b,c,d* are supplied by courtesy of Medway College of Design, and illustrate typical examples of aesthetic industrial models. All models shown are constructed mainly of wood.

Fig. 7.2 shows a full-size clay model of a new car design by the Ford Motor Company. Interior features and external panels are all shaped from a special industrial clay to resemble the finished parts with total accuracy. Modellers use a variety of templates and carving tools to create the required forms. When completed, all dimensions may be recorded, all digitised points

Fig 7.1a Hand drill

Fig 7.1b Car indicator lever

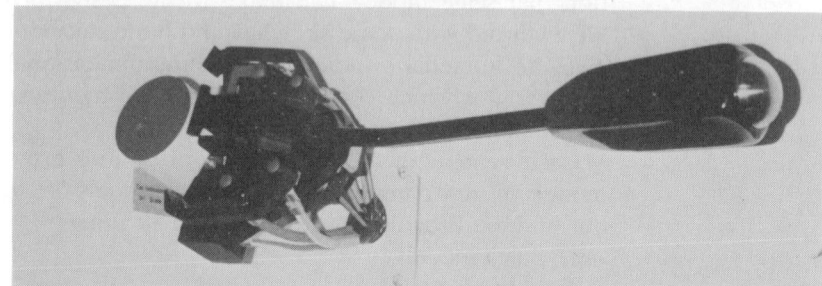

Fig 7.1c Wood plane

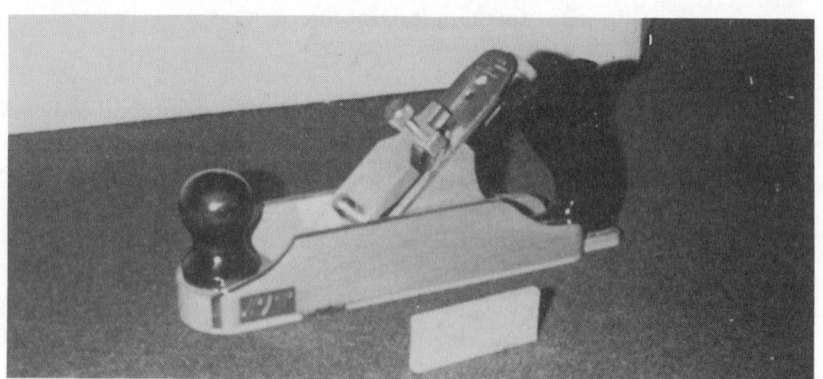

Fig 7.1d Stereo headphones

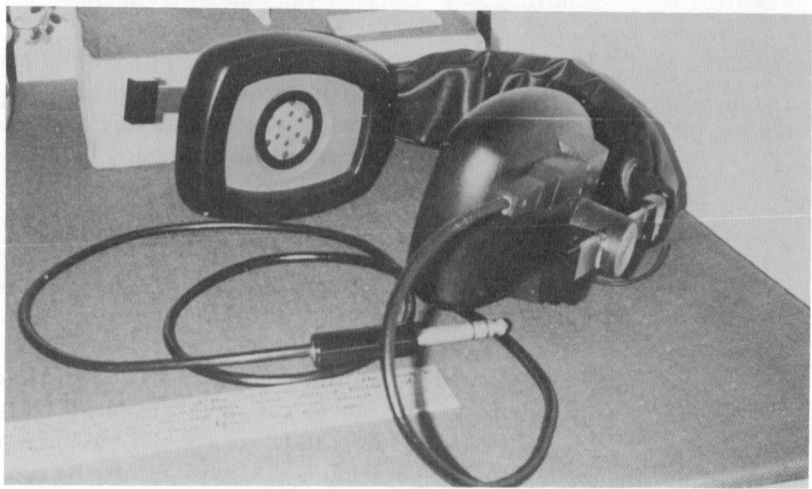

are stored in a computer in three-dimensions, and $\frac{3}{8}$ or $\frac{1}{4}$ scale models are then made for wind tunnel testing.

Aesthetics can be considered as one part of ergonomics. It is thus only logical that other ergonomic aspects are also tested with this type of design model, e.g. the interior of the Ford car model in Fig. 7.3, not only indicates the aesthetic qualities of the product but also the comfort and efficiency of the handwheel and controls.

Fig 7.2 Full-size clay model

Fig 7.3 Model of car interior

4 *Test Models* With increased industrial competition, designers are often required to reduce weight and maintain stiffness of components to such an extent that they become extremely difficult or time-consuming to analyse entirely by calculation procedures. As computer-aided design techniques advance this becomes less of a problem, but it seems likely that scale test models will continue to provide a quick, cheap and efficient means of analysing complicated mechanical designs.

The usual procedure here is that accurate scale models are constructed and tested under expected loading conditions, taking account of material, size ratio and load ratio. Typical applications include aircraft and car models in wind tunnel tests, ships and oil rigs in test tanks, steelwork structures under static and dynamic loads, and photo-elastic analysis of transparent plastic models.

7.2 Qualitative and Quantitative Types of Test Model

Test models may be classed as either *qualitative* type or *quantitative type.*

Qualitative models are essentially visual and may be constructed from fairly flexible material which will aid the designer to see how the structure distorts and deflects under load in comparison with alternative designs.

Quantitative models are used as an alternative to calculation in determining stiffness values, stress levels, vibration characteristics and frictional properties.

Materials for qualitative and quantitative test models should have uniform properties, should obey Hooke's Law, and should not suffer severely from creep. Typical materials include paper, cardboard, balsa wood, perspex, PVC and rubber.

The following examples illustrate typical applications of test models obtained from a series of papers supplied by the National Engineering Laboratory.

1 *Rubber Models of Diesel Engine Bearing Support* (Qualitative)
Figs. 7.4a,b,c, supplied by courtesy of Ruston and Hornsby Ltd., show silicon rubber models of a main bearing support under rapid load test. This test was conducted to eliminate stress in the transverse walls of a crankcase which supported the main bearings. This weakness caused deformation which closed the bearing under firing load and restricted crankshaft rotation. Fig. 7.4a shows the model of the original design, which gave a clear indication of the manner of deformation. The second model (Fig. 7.4b) showed improvements under test by removing two holes in the wall. The third model (Fig. 7.4c) gave the best results under test and indicated the superiority of straight lower walls over the original curved shape under this type of loading.

Photographic record of the deflection patterns was obtained on this test by sprinkling the white models with dark granular material.

Fig 7.4a Original design (of main bearing support)

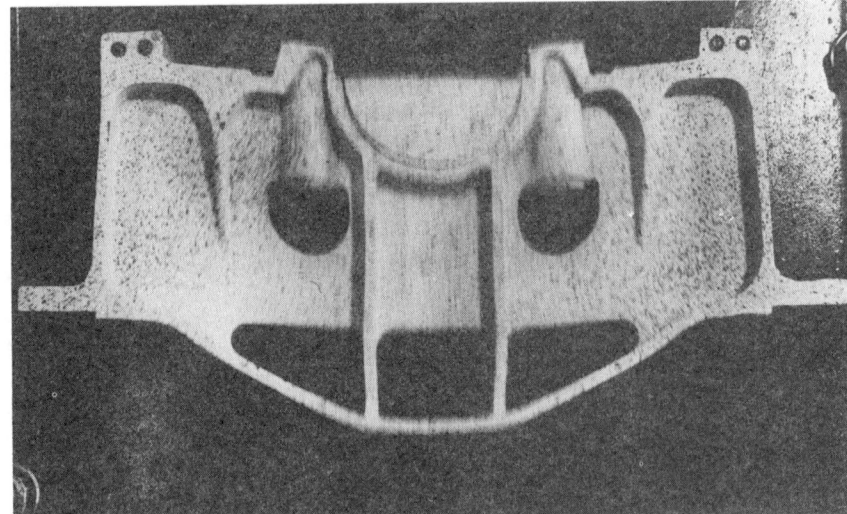

Fig 7.4b Modified design

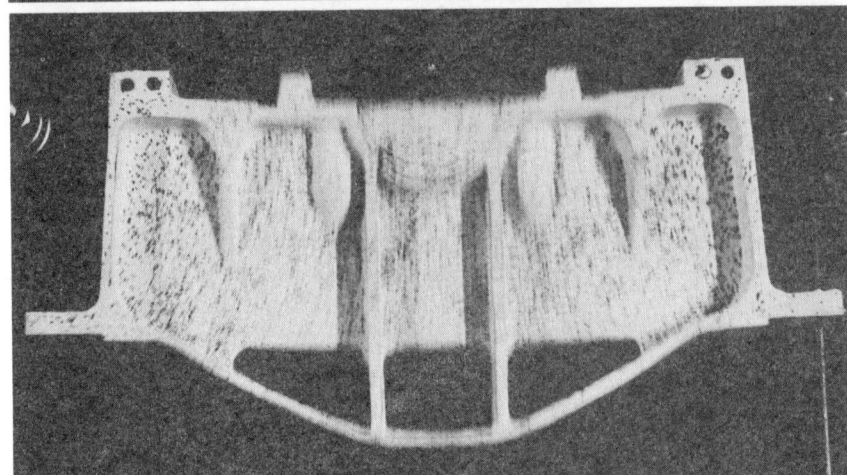

Fig 7.4c Final design

2 *Rubber Models of Lathe Tailstock* (Qualitative)

Fig. 7.5 shows four alternative designs of a tailstock modelled in silicon rubber. Tests similar to that described in example **1** were undertaken to compare the deformation of each design under dynamic loading.

Fig 7.5 Tailstock models

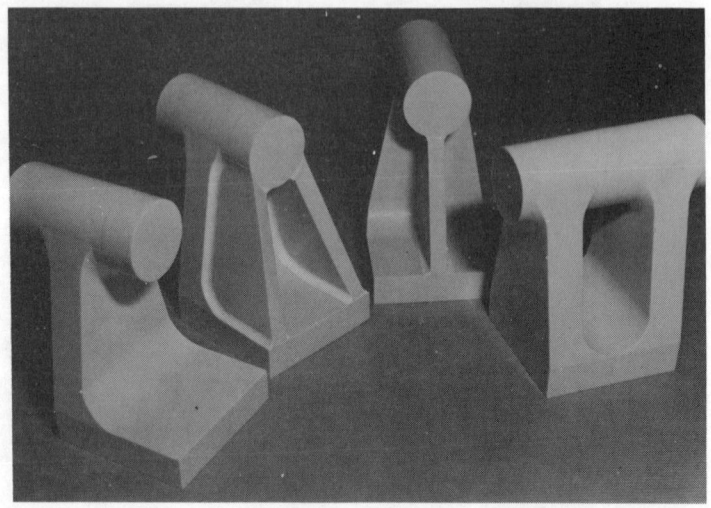

3 *Dockside Crane Structure* (Qualitative)

Fig. 7.6 shows a paper model of the transitional structure between portal legs and upper structure of a dockside crane. This is similar to one of two 1:30 scale models by Vickers Ltd., Elswick Works, made from graph paper and joined with Sellotape. When the two models, with different internal stiffening ribs, were loaded vertically and horizontally on the cylindrical section, it was immediately clear which stiffening method was superior. The paper model exercise which took only two man-days thus proved to be a highly efficient and economical qualitative design comparison technique.

Fig 7.6

4 *Machine Tool Frame* (Qualitative)

Fig. 7.7 shows a test on a foam-plastic model of a machine tool frame section. This was to investigate the effect of a wall opening on the torsional stiffness of the structure. Wall openings are often necessary for access and can reduce torsional stiffness by as much as 50%. The calculations involved can be laborious and models like this are a cheap and quick method of comparing various designs.

Fig 7.7

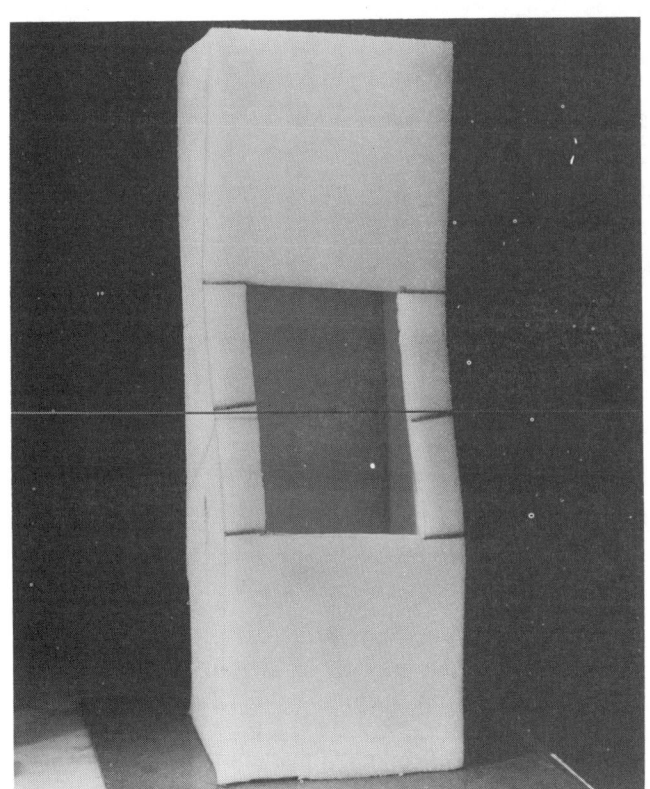

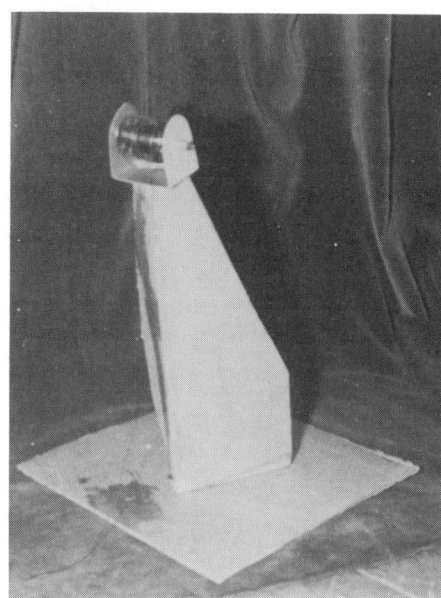

Fig 7.8

5 *Crane Mast* (Quantitative)

Fig. 7.8 shows a 1:12.75 scale cardboard model of a six-metre crane mast similar to one made by Vickers Ltd., Scotswood Works. Here, the designers wished to find the thinnest possible plate whilst maintaining required stiffness. The structure could not be analysed mathematically in the time available. In quantitative exercises like this, a scale test load must be calculated to account for the scale size and model material. After modifications, an acceptable scale deflection was recorded when applying the calculated load at the top of the mast.

Also, *brittle lacquer* was sprayed on the model to locate the high stress points on the mast. On the areas of worst stress, these lacquer-coatings cracked to reveal the stress pattern. Under the scale test load it was thus revealed that the point of maximum stress lay about one third of the way up the nose of the mast.

Fig 7.9

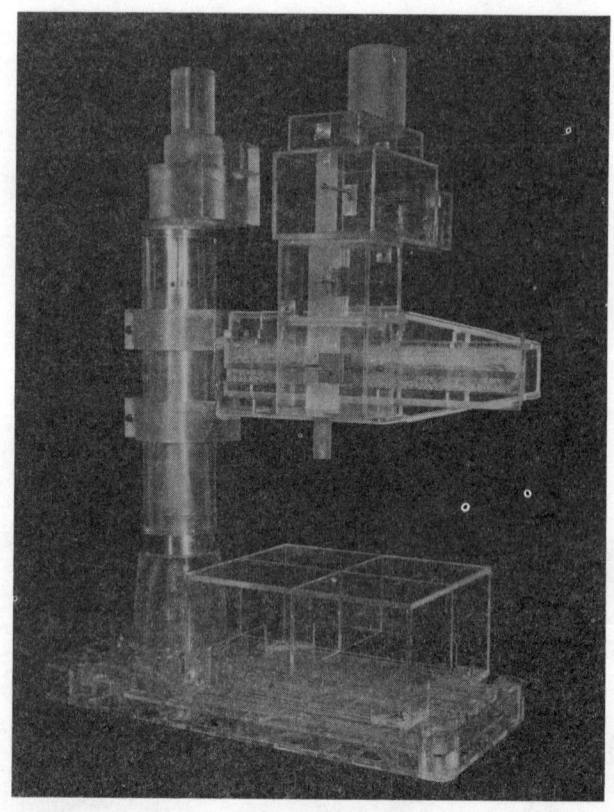

6 *Perspex Model of Radial Drill* (Quantitative)

Fig. 7.9 shows a $\frac{3}{8}$ full size Perspex model of a radial drill used for quantitative testing. From this model the stiffness value of the drill structure was calculated by measuring deflection values while applying scale test loads. Of course, a metal model could have been made but would have been more difficult and expensive to produce, and would have given very small deflections which are difficult to measure accurately.

[Examples **2**, **4** and **6** are based on the work of F. M. Stansfield and colleagues at the Machine Tool Industry Research Association (MTIRA) and are published by kind permission of the Director of MTIRA.]

**7.3
Calculations for
Quantitative
Tests**

Many formulae have been developed, mainly using dimensional analysis, which relate properties of scale model and prototype design. Three of these are listed below:

$$\textit{Deflection} \qquad \delta_m = \delta_p \cdot \frac{F_m}{F_p} \cdot \frac{l_p}{l_m} \cdot \frac{E_p}{E_m}$$

$$\textit{Stress} \qquad \sigma_m = \sigma_p \cdot \frac{F_m}{F_p} \left(\frac{l_p}{l_m}\right)^2$$

$$\text{Natural frequency} \qquad f_m = f_p \cdot \frac{l_p}{l_m} \left(\frac{E_m}{E_p}\right)^{\frac{1}{2}} \left(\frac{\rho_p}{\rho_m}\right)^{\frac{1}{2}}$$

where suffix p = prototype

suffix m = model

δ = deflection (mm)

F = force (N)

E = Young's Modulus (N/mm^2)

l = length (mm)

f = natural frequency (Hz)

ρ = density (kg/m^3)

σ = stress (N/mm^2)

Example 7.1 A steel crane must not deflect by more than 0.25 mm when subjected to its maximum load of 1 MN and an allowable stress of 30 N/mm^2. Using a Perspex scale model of quarter full size, determine whether the crane design is rigid enough for its maximum load application

Take stress limit of Perspex as 5.5 N/mm^2.

E for steel = 200 000 N/mm^2 $\qquad$ E for Perspex = 3100 N/mm^2

Solution First calculate the equivalent test load on the model (F_m value)

$$l_m/l_p = 1/4$$

$$\sigma_m = \sigma_p \frac{F_m}{F_p}\left(\frac{l_p}{l_m}\right)^2 \qquad \text{and therefore}$$

$$F_m = F_p \frac{\sigma_m}{\sigma_p}\left(\frac{l_m}{l_p}\right)^2$$

$$= 1 \times 10^6 \times \frac{5.5}{30}\left(\frac{1}{4}\right)^2 = 11\,460 \text{ N}$$

This test load is then used to calculate the allowable deflection of the model:

$$\delta_m = \delta_p \frac{F_m}{F_p} \cdot \frac{l_p}{l_m} \cdot \frac{E_p}{E_m}$$

$$= 0.25 \times \frac{11\,460}{1 \times 10^6} \times \frac{4}{1} \times \frac{200\,000}{3100}$$

$$= 0.739 \text{ mm}$$

Therefore if the Perspex scale model does not deflect by more than 0.739 mm under a test load of 11 460 N, then the steel crane design is considered adequate for the application.

Vibration analysis

Quantitative models are often used to investigate the natural frequency of vibration of a proposed design.

7.4 Wind Tunnel Testing

The testing of scale models in wind tunnels has become a major contribution to design and development programs particularly in the automobile and aircraft industries. Wind tunnels range from makeshift arrangements adapted from simple ducting and vacuum cleaners or hair driers, to highly powered structures for aircraft testing several kilometres long.

Figs. 7.10a,b shows scale models of cars in wind tunnels tested by the Ford Motor Company. Aerodynamic testing has had a considerable revival in the automobile industry since the western world became conscious of dwindling oil resources and the consequent high cost of fuel. Aerodynamic drag and its adverse effect on fuel consumption is considerably affected by the shape of the vehicle. Car designers use wind tunnels to simulate forward motion by blowing a current of air over a stationary model.

Calibrated scales supporting the wheels measure the reaction of the body. From the test findings a drag coefficient C_d value is determined. Fig. 7.11 shows a diagram, supplied by the Ford Company, of the C_d values of various shapes. A high C_d value will give more wind resistance and is thus likely to have a higher fuel consumption. Also shown is Ford's aerofoil front grille which has been developed in wind tunnels to reduce drag.

Aerodynamic drag is not the only thing which can be established by the wind tunnel testing of car models. The front air dam on the model in Fig.

Fig 7.10a

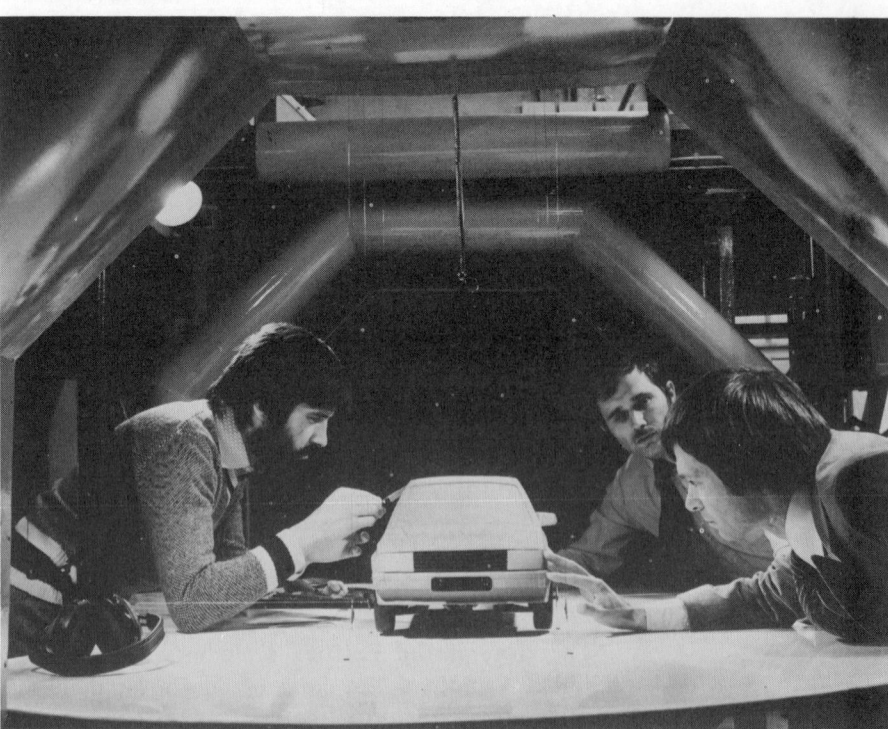

7.10*b* was being developed to reduce aerodynamic lift at high speed which would cause dangerous instability. Also Fig. 7.12 shows Ford diagrams of how dirt deposit on the rear window may be reduced by changing from square-back to semi fast-back design.

Fig 7.10*b*

Fig 7.11

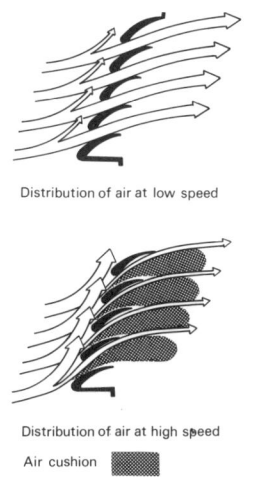

Distribution of air at low speed

Distribution of air at high speed

Air cushion

Ford's aerofoil front grille causes air at high speed to be diverted over the car, to reduce drag.

c_D OF SIMPLE SHAPES

OBJECT		c_D	OBJECT		c_D
Parachute	Air velocity	1·35	Hemisphere	Air velocity	·41
Flat Plate (Square)		1·17	Capri S		·375
Cylinder		35–1·20	Cone (30°)		·34
Cone (60°)		·51	Land Speed Record Car		·11
Sphere		10 – ·50	Wheel-Less Teardrop		·10
1953 Zephyr		·48	Airfoil		·05

Drag coefficients of a variety of shapes, showing where conventional cars fit into the total scene.

The aerodynamic developments in the Escort have now been adopted for many other modern cars. This is clearly seen in the design of the Vauxhall Calibra Turbo, shown in Fig. 7.14.

Fig 7.12

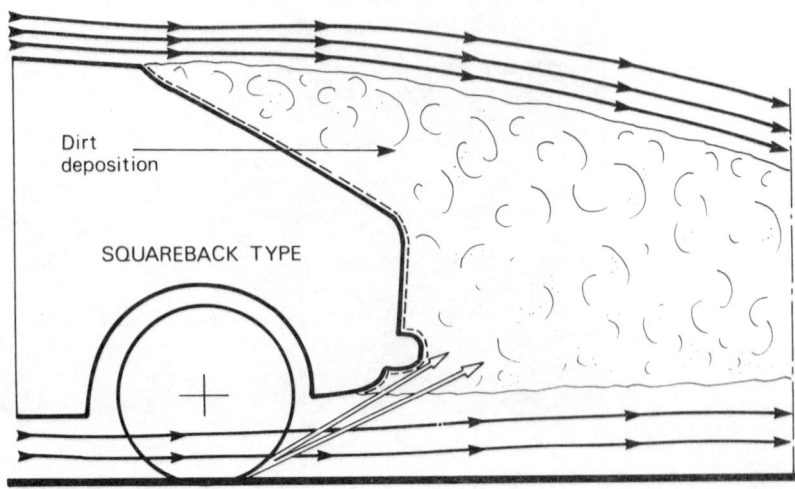

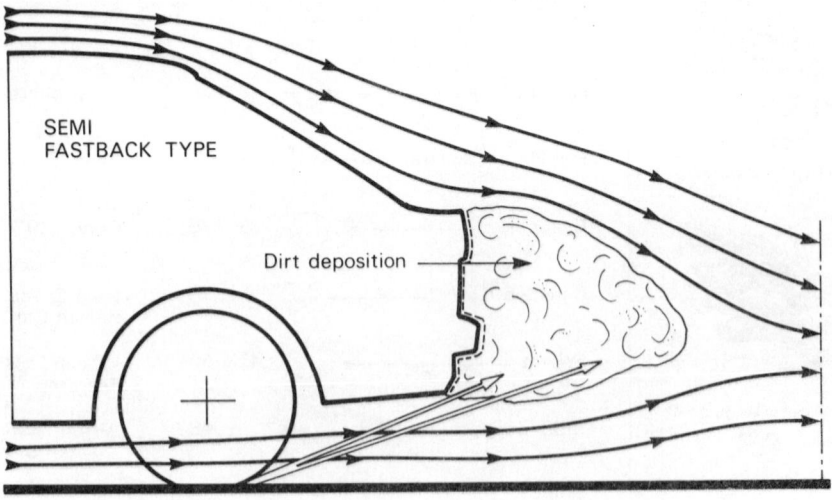

Fig 7.13

Fig 7.14

**7.5
Assignment
Case Studies**

1 *Balsa Wood Bridge Models*

Students (at Canterbury College of Technology) were required to construct simple model bridges of common span and weight and test the models for maximum deflection at mid span. The basic objectives were laid out as follows:

Assignment objectives

1) The students are each required to construct a model bridge of total span 0.5 m from balsa wood and cement. They will use approximately equal amounts of material, but will determine the exact mass in kilogrammes.

2) The students will suspend a 10 N load at mid-span and measure the deflection produced at mid-span by using a dial test indicator. They will thus calculate the stiffness value. They will determine the stiffness/mass ratio.

3) The students will compare the stiffness/mass values and draw conclusions regarding the rigidity of beam structures and its relation to:

    a) Shape of structure

    b) Patterns of framework

    c) Design of joints.

A typical cross-section of results is shown in Table 7.1 and Figs. 7.15*a* to *f*.

Table 7.1

Fig.	*Designer*	*Deflection* (mm)	*Mid-span load* (N)	*Stiffness* (N/mm²)	*Mass* (kg)	*Stiffness/mass ratio*
7.15*a*	Hawkes	0.37	10	10/0.37 = 27	0.06	27/0.06 = 450
7.15*b*	Parry	0.65	10	10/0.65 = 15.4	0.08	15.4/0.08 = 193
7.15*c*	Dean	0.35	10	10/0.35 = 28.6	0.09	286/0.09 = 318
7.15*d*	Saunders	1.35	10	10/1.35 = 7.4	0.07	7.4/0.07 = 106
7.15*e*	Kirk	0.38	10	10/0.38 = 26.3	0.05	26.3/0.05 = 526
7.15*f*	Ullyett	0.58	10	10/0.58 = 17.2	0.08	17.2/0.08 = 215

Fig 7.15   a

b

c

d

e

f

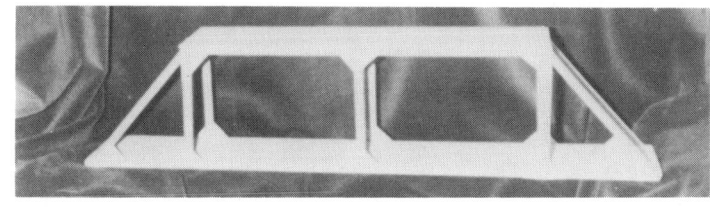

2 *Simple Wind Tunnel*

Fig. 7.16 shows a basic wind tunnel of about 1 metre length constructed by students at Canterbury College of Technology. It has a wood base and mainframe and a transparent acetate film body and metal rods inside to support scale models. The right-hand end cap is a plastic funnel which may be detached to accommodate the models.

The wind effect was obtained by attaching a vacuum cleaner hose to the plastic funnel. The left-hand end is filled with hollow plastic straws to allow the wind flow and accommodate glass tube smoke jets.

Various shapes of scale model were tested in the tunnel and the smoke flow patterns observed and analysed. Comparisons of the aerodynamic characteristics in each shape of model were then made.

Fig 7.16*a* Wind tunnel

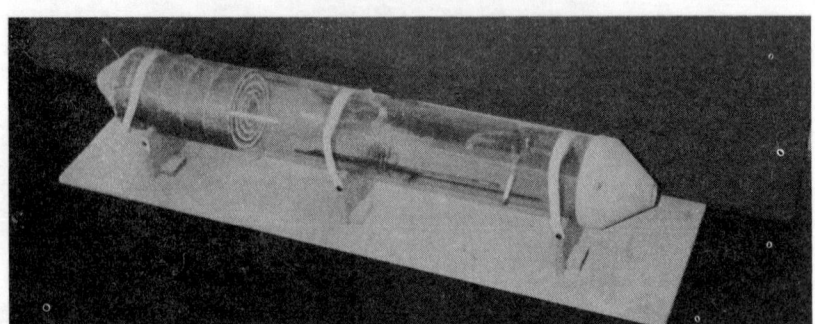

Fig 7.16*b* Exploded view of wind tunnel

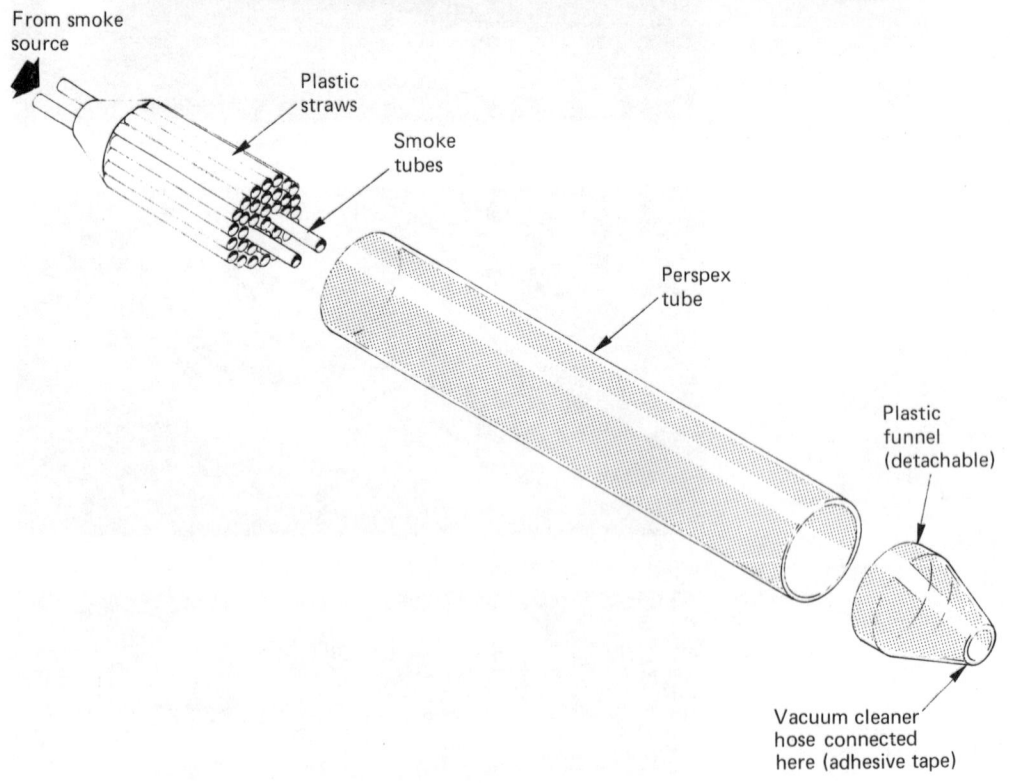

Problems

7.1 State the main types of scale model used in engineering design and describe their uses.

7.2 Distinguish between the terms qualitative test models and quantitative test models. Describe some applications of each.

7.3 Describe some common applications and methods of wind tunnel testing.

7.4 Referring to Assignment Case Study 1 on page 140 conduct a similar exercise for a balsa wood model of a crane mast to the approximate outside dimensions shown in Fig. 7.17. (This problem could form part of Assignment 7 on page 213.)

Fig 7.17

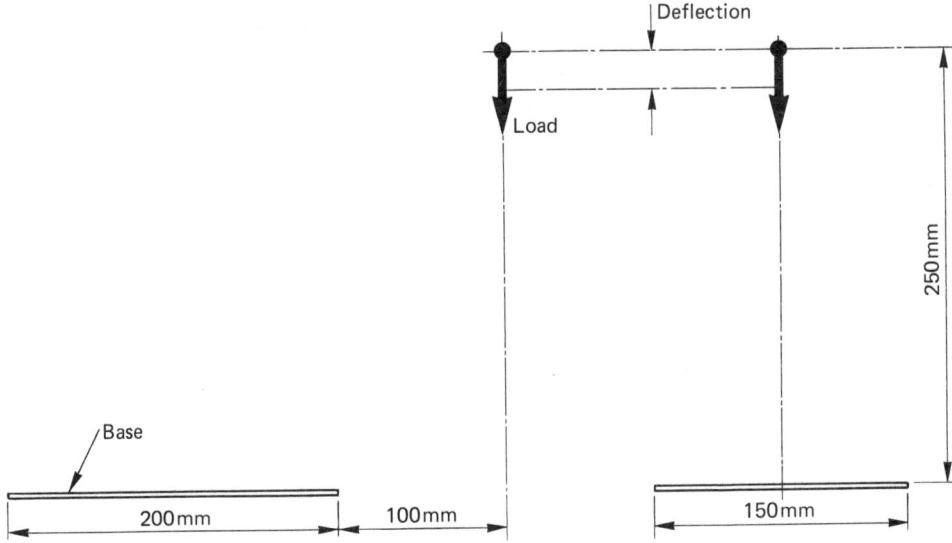

7.5 *a*) Construct a wind tunnel similar to that shown in Assignment Case Study 2 on page 142.

b) Construct qualitative scale models of a car or airplane design to approximately the same outside dimensions and surface area. (Plasticine would be a suitable material here.)

c) Compare the aerodynamic characteristics of the designs.

7.6 *a*) Referring to problem **7.4** construct this model as a quantitative model in Perspex.

b) Applying a load of 10 N, measure the deflection at the top of the jib.

c) Referring to worked example 7.1 (p. 135) and using the same values of allowable stress and Young's Modulus, determine the maximum load and expected deflection if your model is assumed to be 1/10 full size of an equivalent steel crane.

8 Aesthetics

8.1 Aesthetics, Ergonomics and Design

Aesthetics is defined in the Oxford Dictionary as: a set of principles of good taste and appreciation of beauty. Although function, costs, safety and the physical aspects of ergonomics may be paramount in the initial stages of the design process, the appearance of a product is often a major factor in its saleability. In many respects this is justified and logical. For example, an untidy design could indicate to a potential customer that the company and its products were inefficient. Similarly, an "old-fashioned" looking design could give the impression of being out of date in its technology.

An engineering design with good aesthetics will thus be pleasing to the eye and give a visual impression of functioning efficiently.

Aesthetic Ergonomics

As defined in Chapter 6, ergonomics is the scientific study of the relationship between people and their working environment. Since the appearance of a design directly affects the relationship between product and user, and can often aid the efficiency of an operation, aesthetic engineering may be considered as an important part of ergonomics.

Industrial Design

This is the name given to that branch of design which deals mainly with the aesthetic and ergonomic aspects of product development.

Industrial design may be undertaken as one of the many duties of the engineering designer, or it may be tendered to a professional consultancy whose specific task is to improve the look and feel of functionally adequate designs.

8.2 Aspects of Aesthetics

Various aspects of aesthetic design will next be discussed, but they should not be considered as rigid disciplines. Ideally they will be inter-related and geared to the functional, ergonomic and manufacturing requirements of the design.

Symmetry and Balance

Nature designed most of its life forms in approximate **symmetry** about at least one axis. This was first copied by architects who, in their turn, were copied by engineers. The human eye is thus conditioned to see designs in symmetrical form and tends to reject asymmetrical shapes as ugly or illogical. Of course, symmetry is favoured by many production processes but is often deliberately adhered to for the sake of appearance.

Human taste may be conditioned to accept asymmetry where it suits function, but even so, in these cases, it still prefers to see a sense of **balance**. Fig. 8.1 shows three arrangements of a control panel. Arrangement (a) has been made unnecessarily symmetrical and is ergonomically poor; arrange-

Fig 8.1 Types of
control panel

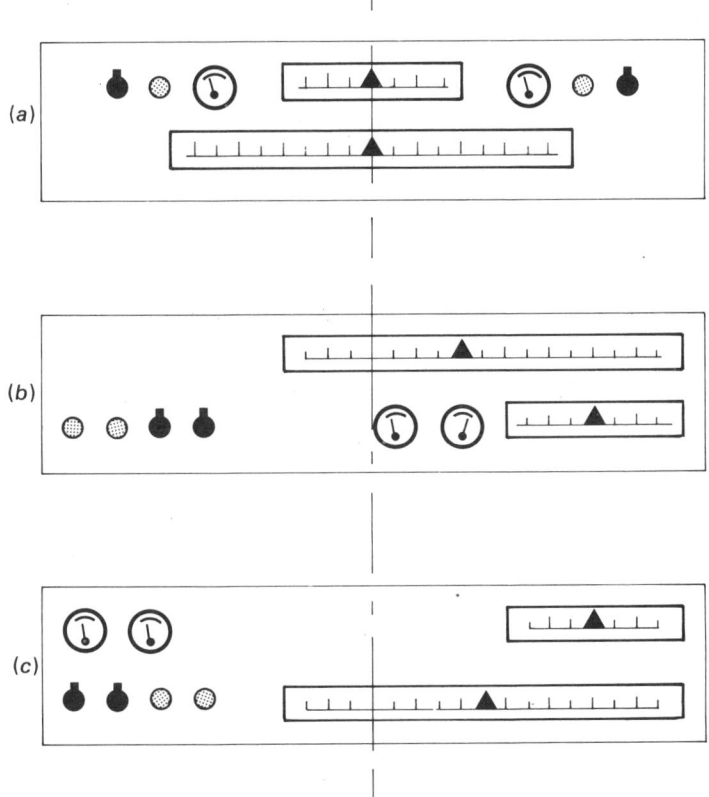

ment (b) employs good ergonomics, but looks unbalanced because the bulk
of the display "mass" is towards the right of the panel; arrangement (c)
maintains good ergonomics whilst obtaining a sense of balance by equating
"moments of mass" about a central datum. An extra mass moment effect
has been added to the left-hand side of (c) by using a darker colour on the
controls.

Continuity

A design which has good **continuity** will appear to form a harmonious
pattern in its profile and arrangement of elements. The expression is thus
associated with the "order", or "tidiness", of a product.

Fig. 8.2 shows a good example of improved continuity in a product. The
before and after views of the re-designed Karpark meter indicates how
aesthetic appeal and functional efficiency may be increased by tidying up
interruptions of contour in display, control and outside profile to give an
"in-line" form to these features. This form also improves order by grouping
the features as separate components in a set pattern.

The Toyota Previa "space wagon", shown in Fig. 8.3 reflects a current social
trend and combines functional design with good aesthetics. Excellent continuity
is achieved by arranging the inner elements as part of the same pleasing pattern
as the aerodynamic outside profile.

Fig 8.2 Karpark meters [*Karpark Ltd.*]

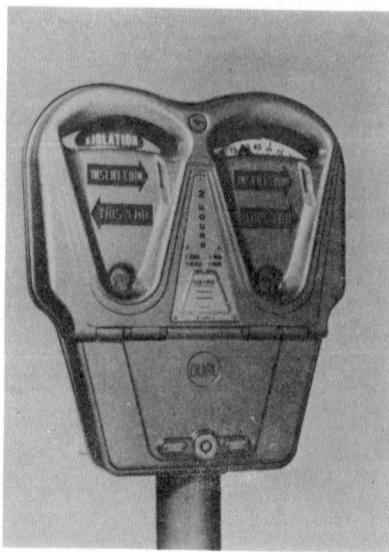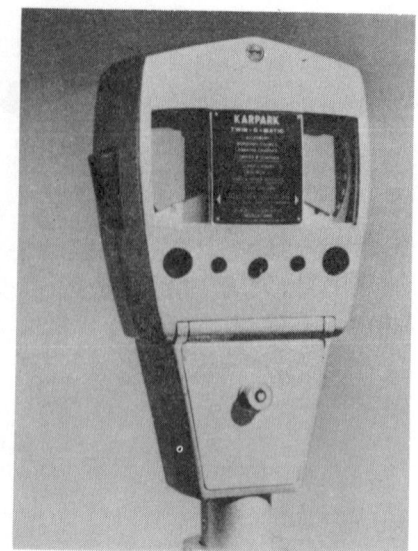

Fig. 8.3 Toyota Previa

The advantages of standardisation have been outlined in Chapter 3 and can also play an important part in aesthetics. Wherever function permits, continuity may be improved by ensuring that component features such as fastenings, displays, controls, bosses, fillet radii, shafts, welds, and structural sections are either identical, or of the same form.

Variety

Whilst standardisation is an important aspect of the design form, excessive uniformity can become tedious to the eye and may result in poor ergonomics.

Variety is particularly useful when marketing ranges of products. The Rover car group once described one of their popular products as "not the car for Mr. Average" and thus gave the impression of relieving the motorist from the boredom of a standard shape of vehicle (see Fig. 8.4).

Fig 8.4

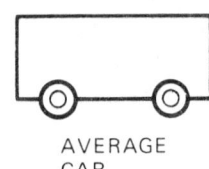

AVERAGE
CAR

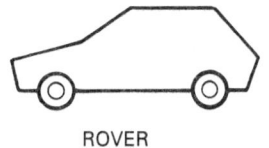

AVERAGE
CAR

ROVER
DESIGN

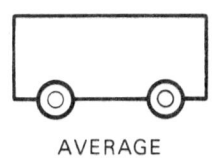

AVERAGE
CAR

The vast choice of threaded fastening types available, most of which are functionally identical, gives another example of the demand for variety in product ranges.

Proportion

Proportion is concerned with the relationship in size between connected items, or parts of items.

Consider the components shown in Fig. 8.5. These items all look strange because they are "out-of-proportion" in terms of their expected shape. Traditionally expected proportions are nearly always developed from sound functional requirements, but can sometimes override the functional aspect. The spanner design in Fig. 8.5 may function perfectly satisfactorily and suit manufacturing processes, but may be rejected as looking out of proportion.

Proportion is thus closely associated with the impression of purpose given by the shape of a component—an aspect which is later discussed in more detail.

However, the designer should not conform blindly to strongly-held stereotypes. Reluctance to change from familiar shapes may discourage design innovation and creativity in styling.

Contrast

Contrast may be defined as the balance of adjacent elements which have clearly different characteristics.

It is interesting to note that the re-designed Karpark meter in Fig. 8.2, in addition to having good continuity, also gave pleasing contrast between its three main features (display, control and outer casing).

Fig 8.5

UNEXPECTED PROPORTION

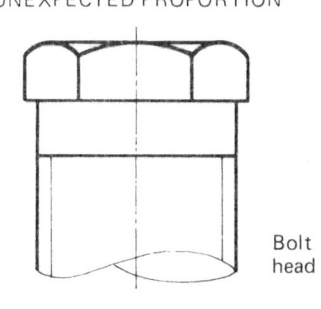

Bolt
head

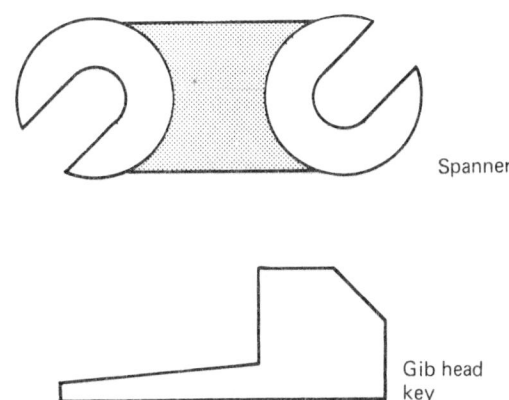

Spanner

Gib head
key

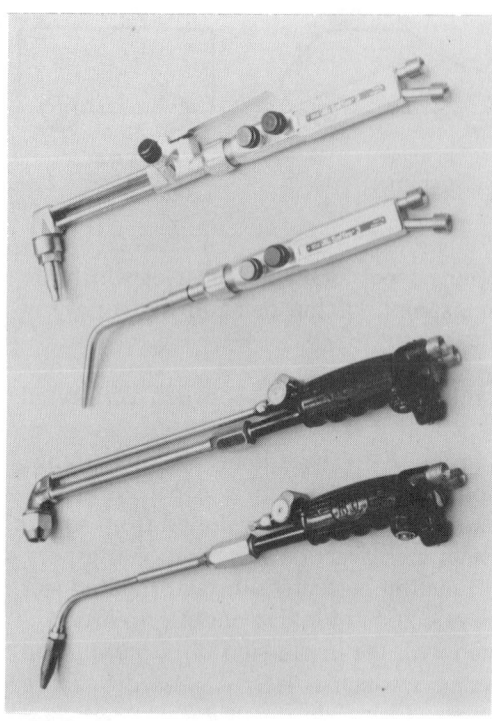

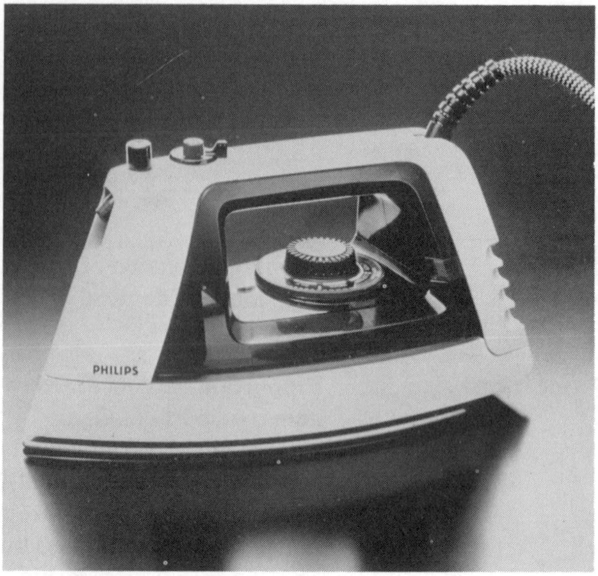

Fig 8.7 Steam iron
[*Philips Electronics*]

Fig 8.6 Welding and cutting outfits [*B.O.C. Ltd.*]

Choice of colours can be particularly important. Fig. 8.6 shows a before and after re-design of welding and cutting outfits to improve continuity and provide a more distinct contrast between features. The extent of colour contrast (reasonably apparent from the black and white reproduction) is not so severe as to be irritating, but gives an appealing blend of foregrounds and background.

Fig. 8.7 shows a design of steam iron produced by Philips Electronics. This is another example of combining good continuity of profile with pleasing colour contrast of elements.

The Impression of Purpose

The aesthetics of engineering must inevitably invoke not only the question "Does it look nice?" but also "Does it look as if it will work?".

The importance of proportion has already been outlined. Shape, generally, can affect the impression of purpose in a design. Strength and stability are often implied by using tapered edges and lines. For example, the levers shown in Fig. 8.8 can be made to look more rigid by using tapers, even though less material is used in the latter case.

The increased strength in bending implied with a tapered section is used a great deal with cantilever beams, as shown in Fig. 8.9.

Stability may be implied by tapering to a wide base (thus lowering the centre of gravity and apparently giving more stable equilibrium) as shown in Fig. 8.10.

Colour contrast may also be used to imply strength and stability, as can reasonably be deduced from Fig. 8.11.

Fig 8.8

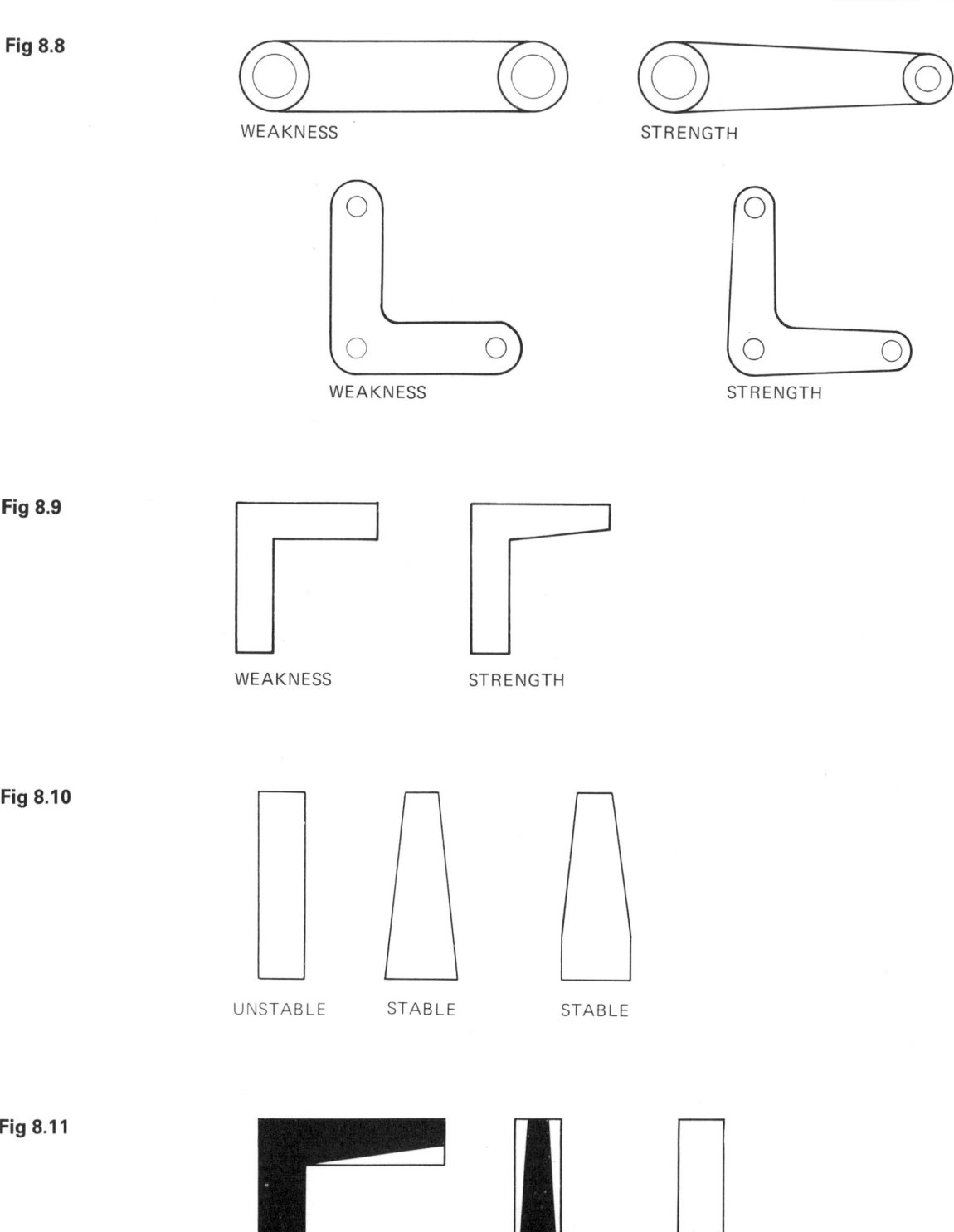

WEAKNESS

STRENGTH

WEAKNESS

STRENGTH

Fig 8.9

WEAKNESS

STRENGTH

Fig 8.10

UNSTABLE

STABLE

STABLE

Fig 8.11

Fig 8.12 Foodmixer
[*Kenwood* Chef]

The Kenwood Chef Foodmixer shown in Fig. 8.12 is a typical example for this discussion. The taper and large radius on the vertical strut implies both strength and stability. The horizontal cantilever, although having a uniform cross-section, appears to be tapered by curving to a slant, and thus implies strength in bending. The latter effect is enhanced using colour contrast.

Mobility may also be implied with shape. Knowledge of aerodynamic design has resulted in tapered and flowing shapes appearing to give higher speeds than rectangular shapes, as shown in Fig. 8.13.

Again colour contrast may be used for mobility, as in Fig. 8.14, and may also improve continuity. (See also Fig. 8.3.)

An impression of precision is often attempted and is obviously essential in instrument design. This is often achieved with the use of clearly defined contours, and continuity of shape, and by avoiding irritating protrusions of profile. Precision is also included as an aspect of style, which is discussed in the next section.

Fig 8.13

SLOW   FAST   FAST

Fig 8.14

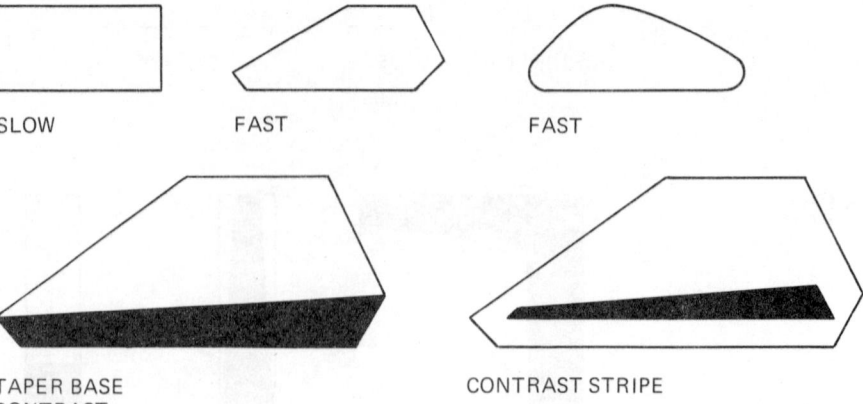

TAPER BASE
CONTRAST

CONTRAST STRIPE

Fig 8.15 Victorian machine frame

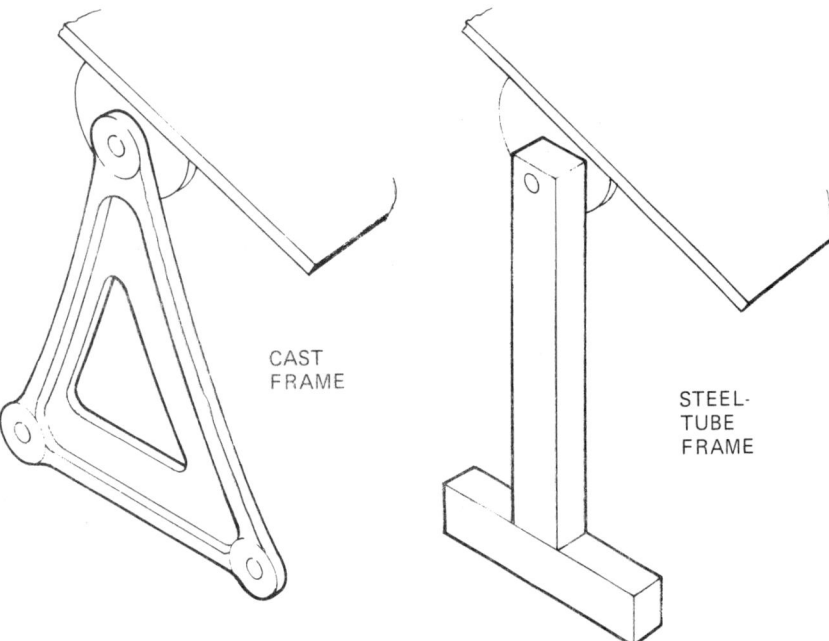

CAST FRAME

STEEL-TUBE FRAME

Fig 8.16

8.3 Style

Style is difficult to define precisely, but it is concerned with a certain visual quality which characterises a design or "sets it apart from the rest".

Good styling will skilfully reflect a current public mood which may be influenced by developments in technology, or by a prevailing social or environmental climate. Good styles set trends and so create new fashions. These, in turn, influence the styles of other products. As previously indicated, engineering design originally followed the fashions of architecture. The Victorian machine frame shown in Fig. 8.15 tends to copy the decorative curves of cast iron fashion which was typical in the pillars, archways, lamposts, and fireplaces of this period.

The influence of new materials and production techniques is particularly significant in the development of new styles. Cast iron is a medium which flows easily into intricate moulds and thus lends itself to decorative curves and protrusions. Steel, however, is more suited to rolling and drawing, and thus the rectangular styles provided by angle, channel, plate and tube section began to emerge.

Fig. 8.16 contrasts a Victorian cast iron drawing board frame with a modern style in steel tube. The rectangular form has many additional advantages which tend to suit the modern environment—including ease of installation, packaging and cleaning. It may also be observed that, although cast iron is still widely used in engineering, it now follows the rectangular form of steel sections as shown in Fig. 8.17

The versatility of form offered by the stronger plastics has made an inevitable impact on the style of modern products. The plastic structure of the modern sewing machine (Fig. 8.18) typifies the unchanged trend from

Fig 8.17

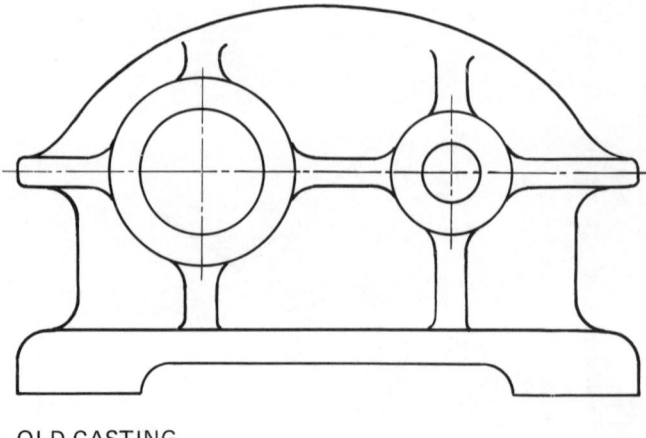

OLD CASTING

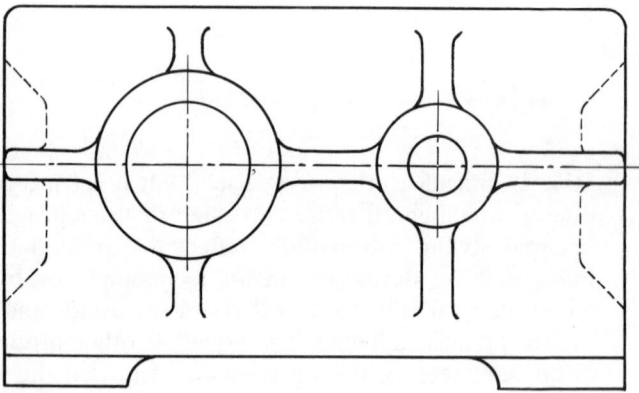

NEW CASTING

decorative protrusions towards smooth continuity, but also a return to curvaceous contours. Another move back to curved profiles is the increased use of lightweight cold-formed steel and aluminium tube and plate. The large bend radii of these sections have a modern appearance which blends particularly well with plastic profiles. A typical example here is the lawn mower produced by Flymo Ltd., shown in Fig. 8.19.

Advances in technology affect not only the form of the particular designs concerned but also often influence totally unrelated products. For example, the aerodynamic development in the shape of aircraft, cars and boats has been copied in radios, vacuum cleaners, computers and typewriters. Fig. 8.20 shows a computer with a distinctly aerodynamic profile. Such shapes give the impression of supplying the technology and precision required to cope with the pace of modern lifestyle. Also, the advances in electronics and instrumentation have left their mark on modern types of stereo systems,

Fig 8.18 Sewing machine [*Frister & Rossmann*]

Fig 8.19 Lawn mower [*Flymo*]

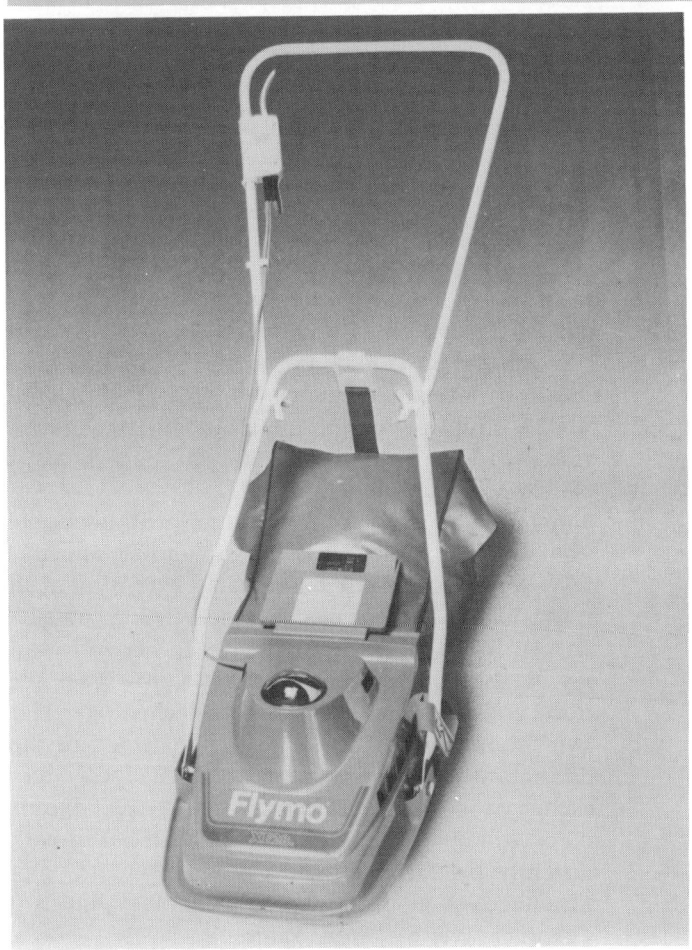

Fig 8.20
Microcomputer
[*Engineering
Computer Services*]

Fig 8.21 The Darby
wood-burning stove
[*Coalbrookdale Co.*]

watches, cameras, and pocket calculators. These often incorporate excessive quantities of displays and controls which serve little function, but, again, give an impression of high technology and precision.

Social and environmental factors such as safety, pollution, and conservation also have an influence on style. For example, the revival of curved profiles in plastics and metal tubing may have been affected by their widespread use in domestic and childrens items where the sharp edges of box shapes could appear unsafe.

Finally, the style of a product will undoubtedly be affected by the requirements of the market at which it is aimed. It is widely believed that many products are purchased as an extension of the customer's personality, in which case efficient market research is essential. Styles aimed at a teenage market, for example, are likely to be biased towards precision and technology. At the other extreme, there is a growing nostalgia market as a rejection of the consumer aspect of modern technology. The desire in some people to own "something made to last" is reflected in the successful marketing of efficient wood-burning stoves which look like old-fashioned cast-iron kitchen ranges, and in the recent revival of the much-loved Morris Minor.

For example, the Darby wood-burning stove shown in Fig. 8.21 and currently produced by the Coalbrookdale Company, incorporates all the advantages of modern heating technology, but is purposely designed with a Victorian-style cast-iron frame.

**8.4
Some Case
Studies in
Aesthetic Design**

1 *Domestic Kettle.*

Figs. 8.22*a,b* show a comparison between a Victorian kettle and modern designs produced by Haden.

The style for each period is distinctly affected by developments in material and manufacture. The shape of the Victorian kettle is as suited to its wrought iron manufacture as is the shape of the moulded plastic in the modern versions. The size of the plastic main body handle and lid handle is ergonomically superior on the modern kettles, since the heat insulation of plastic allows a larger section than the metal equivalent of the older kettle.

Other attributes of the modern designs are: continuity of profile between main handle and body; stability of body on electrical stand, pleasing proportion between body, handle and spout; pleasing colour contrast between body and stand.

The overall comparison on style is that the Victorian kettle looks decorative and impractical, whilst the Haden designs imply efficiency and precision for the modern environment.

Fig 8.22*a* Victorian kettle

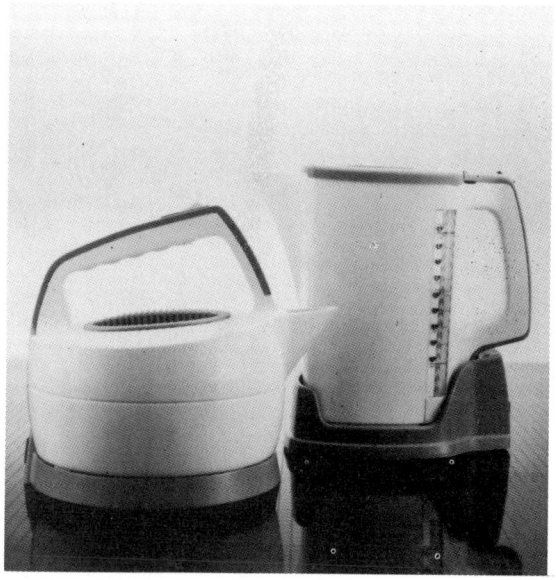

Fig 8.22*b* Automatic kettles [Haden]

2 *Centre Lathe*

Figs. 8.23*a,b* compare a modern lathe produced by Mondiale with an older design. The old design has a slightly Victorian appearance and contains a great deal of "visual clutter", particularly in its controls. The up-dated design is a good example of how purpose is enhanced by the removal of visual clutter and the clarification of basic elements. The continuity of profile and controls, clear statement of working parts, efficient contrast of controls and main structural features, and good sense of stability and proportion, all help to give the modern lathe a superior impression of efficiency and precision.

Fig 8.23*a* "Gallic"
Lathe [*Mondiale*]

Fig 8.23*b* Redesigned
"Gallic" lathe
[*Mondiale*]

3 *Car Interior*

Fig. 8.24 shows a side view of a Ford Sierra 4 X 4 car interior. This is an excellent example of how new materials and processes affect styling. Most of the visible items here are made from modern plastics, and the accompanying trend towards smooth curvaceous contours is almost totally adhered to (see also Fig. 6.16 for anthropometric aspects).

Fig 8.24 Interior of the
Sierra 4 X 4

4 *CNC Milling Machine*

Fig. 8.25 show typical examples of how new technologies have affected the appearance of machine tools. The Series I Interact CNC (computer numerically controlled) milling machine from Bridgeport Textron is characterised by stable continuity of base structure, clear definition of functional features, and pleasing appearance of controls and displays. The computer control unit (*b*) blends efficient colour contrast with a neat continuous profile. The joystick and electronic handwheel control in (*c*) makes use of excellent colour contrast. The milling machine as a whole involves, of course, many anthropometric considerations, some of which have been discussed in Chapter 6.

Fig 8.25 CNC milling machine, showing close-ups of *b*) the computer control unit and *c*) the joystick and electronic hand wheel control [*Bridgeport Textron*]

a

b

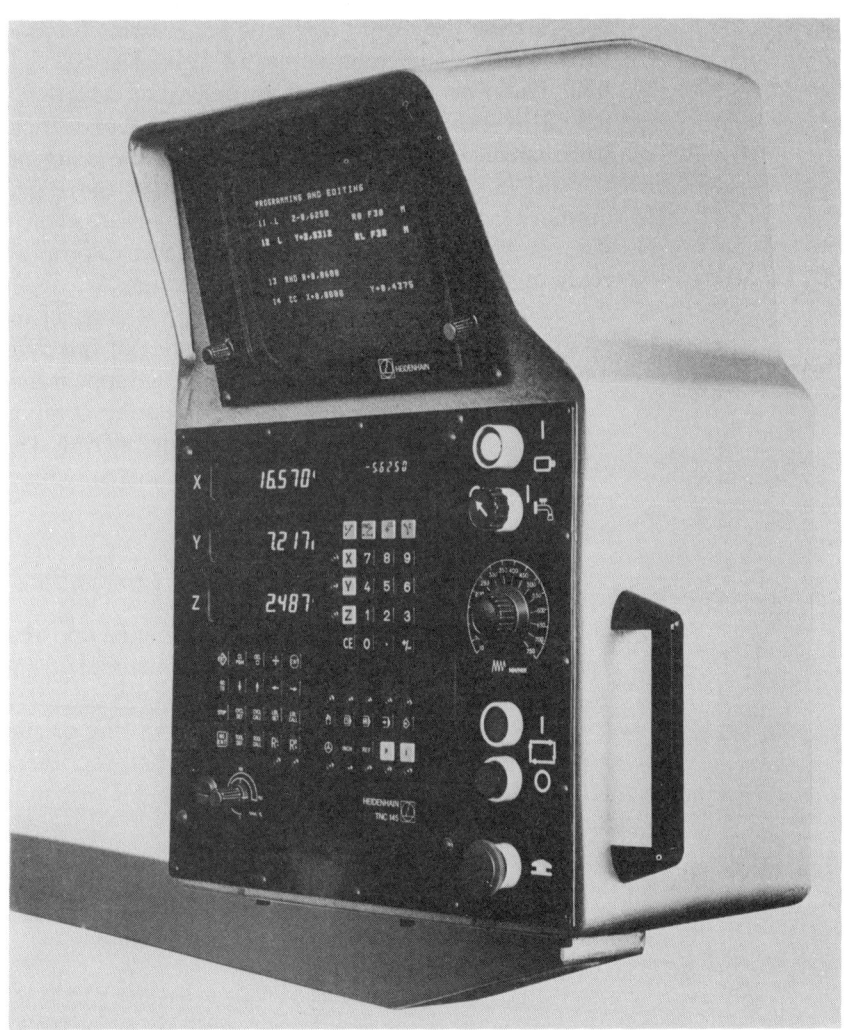

c

Problems

8.1 Define the term "aesthetic engineering" and explain its importance to the saleability and ergonomics of a product.

8.2 Describe the duties of an industrial designer.

8.3 List some practical advantages to a symmetrical shape of product. Give some specific examples of where symmetry is not possible and suggest, with the aid of neat sketches, ways in which these products may retain good aesthetics.

8.4 Define the terms: continuity, variety, proportion, and contrast, with particular reference to engineering design.

8.5 For each of the terms listed in **8.4** choose one product in which you observe deficiences. Make neat sketches of improved alternatives.

8.6 Describe methods of enhancing the appearance of purpose in product design. List typical examples in engineering components.

8.7 Give specific examples of how engineering designs may be affected by changes in style. Describe some important influences on styles of fashion in engineering.

8.8 Investigate an old design of
  a) Shaping machine
  b) Bench drill
in a workshop with which you are familiar and re-design the machines to improve aesthetic features in a similar manner to the case studies described on pages 155–159.

Follow initial sketches by a formal drawing of each re-design and list the aesthetic improvements made in a written report.

Bibliography
W H Mayall *Industrial Design for Engineers*
  (Newnes-Butterworth).
W H Mayall *Principles in Design* (Design Council, London).
Eskild Tjalve *A Short Course in Industrial Design*
  (Newnes-Butterworth).

9 Computed-aided Drawing and Design

Computers are being used increasingly for both design and detailing of engineering components in the drawing office. The creation of engineering drawings using a CADD system offers a manufacturer the following advantages:

a) Uniform design standards.
b) Consistent specification of components.
c) Elimination of inaccuracies caused by hand-copying of drawings and inconsistency between drawings.
d) Easier modification to drawings.
e) Simpler production of similar drawings having minor changes.
f) Increased productivity.

To the designer, the CADD system becomes the sketch pad, allowing the facility to draw the basic design, evaluate and modify very quickly. To the draughtsman, repetitive work is eliminated, and it allows concentration on improving standards and styles of drawing.

The rate at which drawings are produced increases. It has been estimated that using conventional draughting techniques a detail draughtsman produces 250 drawings annually. About 50 of these are major drawings occupying about half the time. Using a CADD system it has been shown that an improvement of over $3\frac{1}{2}$ times this work output can be achieved. This value, of course, depends on the nature of the product.

The main enemy of most draughtsmen is repetitive work and the resulting boredom. This results in drawing errors and the need to modify drawings. With a CADD system the draughtsman is continually learning new techniques on the system. Attention is held more acutely than it is with conventional draughting methods. Consequently, interest is held and, hence, productivity goes up.

In many cases, during the construction of a drawing using conventional drawing board and pencil, it is necessary to erase parts should a mistake or change of mind occur. With a CADD system it is very easy, for example, to erase a line. This is done with the aid of a digitiser by simply locating the origin of the line required to be removed. There are, of course, no messy marks left on the drawing after this process and no indication that the offending line had ever been there.

As well as engineering drawings, the CADD system can produce parts lists, material requirements and planning charts, etc. It can also be used to perform basic calculations such as areas and second moments of area, work out centres of gravity, do geometric calculations and carry out stress analysis using finite element techniques. These calculations are done during normal access to the drawing with little effort and almost instantaneous response.

Because it is easy to produce design variants and store them, it is possible to analyse the structural and functional aspects of any design in great detail using techniques just mentioned. This can save a great deal of money in developing a product, and on some systems it is possible to rotate machine parts and simulate their actual operation when put into service.

The CADD system therefore is a versatile tool, able to assist the designer and draughtsman to improve their work output, remove repetitive boring work, and give consistent, quality drawings.

9.2
The Complete
CADD System

The central processor unit (CPU) of the computer is the heart of the CADD system. In computer terms this is the *hardware* necessary to perform and control all operations in producing the drawing. For the computer to carry out these operations, it is necessary to feed information to it in the form of a sequence of instructions, called a program, and placing it in the computer memory. These programs are called the *software*.

Connected to the computer are *workstations* and other attachments called *shared peripherals*. Fig. 9.1 shows a typical system consisting of workstations, disc storage unit, punch tape unit and hard copy unit (*plotter*), etc.

The CADD Workstation

A typical workstation could consist of
a) A graphics screen or visual display unit (VDU).
b) An alpha-numeric message display (word and number screen).
c) A command tablet.
d) A keyboard.
e) A cursor control device (e.g. joystick, light pen, digitising pen-and-tablet).
f) A mini plotter unit.

Fig. 9.2 shows some common arrangements of CADD workstations (the mini plotter could be added to any of these).

The VDU is used for projecting the drawing or design such that the draughtsmen or designer can produce completed working drawings. The displayed drawings when completed are small in relation to their real size and have to be enlarged on the screen when developing its shape and form.

The workstation needs to look right and have aesthetic appeal. It also needs to be of good ergonomic design, for the designer/draughtsmen will spend many hours in this environment and therefore all controls must be within easy reach. In order to achieve this, the chair should be regarded as part of the system, which will help prevent premature tiredness.

The keyboard is used for both programming the computer and instructing the computer to perform drawing and design functions. It consists basically of a typewriter keyboard which allows access to the computer. Instructions which are typed in appear on the VDU screen so that a check can be made that the correct information is being put into the computer.

The mini plotter is not an essential part of the system, but is useful in printing out whatever is on the VDU screen for use as a reference—for

Fig 9.1 CADD system

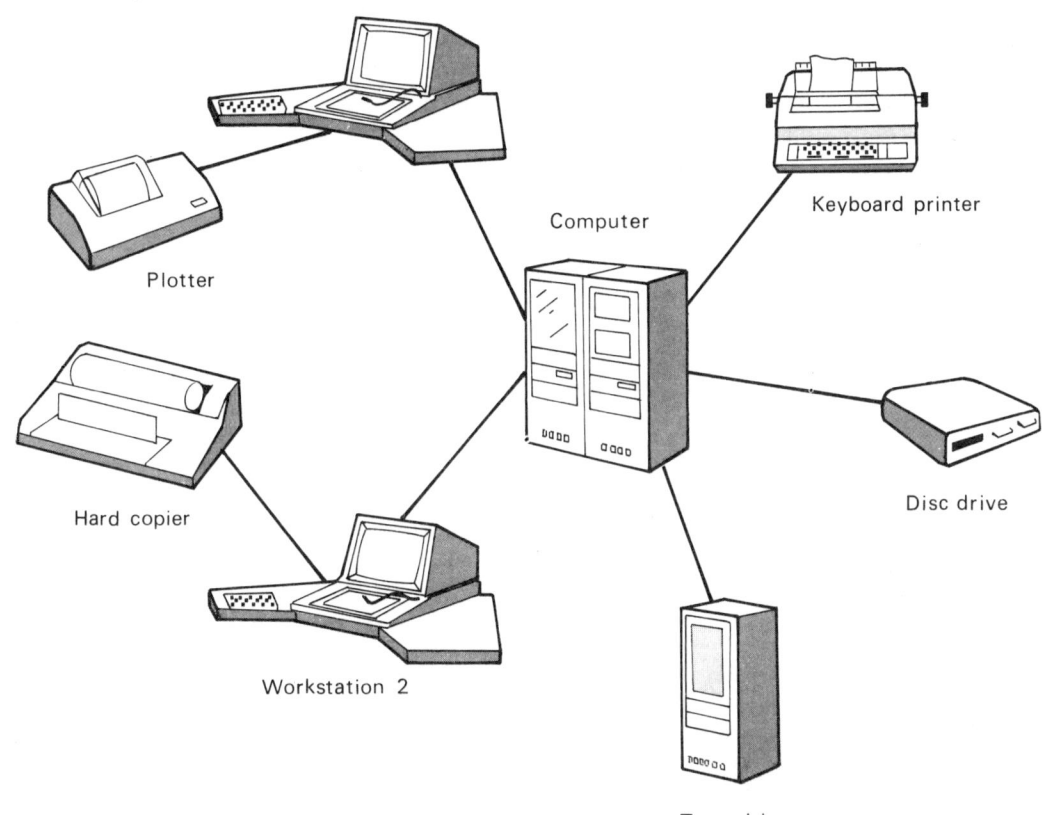

Workstation 1

Plotter

Hard copier

Workstation 2

Computer

Keyboard printer

Disc drive

Tape drive

Fig 9.2 Typical
arrangements of
CADD workstations

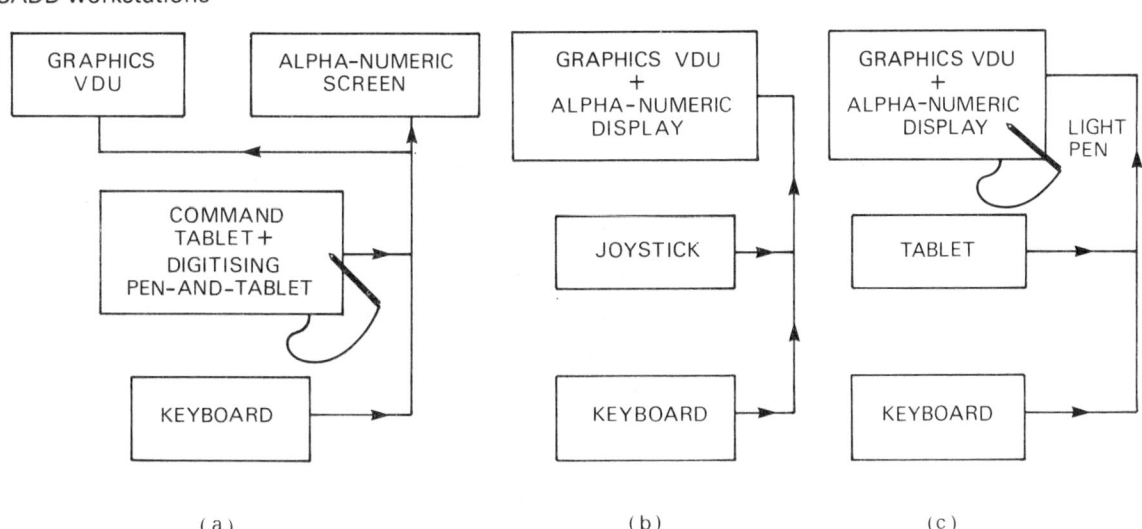

GRAPHICS VDU	ALPHA-NUMERIC SCREEN

COMMAND TABLET+ DIGITISING PEN-AND-TABLET

KEYBOARD

(a)

GRAPHICS VDU + ALPHA-NUMERIC DISPLAY

JOYSTICK

KEYBOARD

(b)

GRAPHICS VDU + ALPHA-NUMERIC DISPLAY

LIGHT PEN

TABLET

KEYBOARD

(c)

example, to reproduce a hydraulic or electronic circuit diagram. There would be no sense in reproducing a copy from the main plotter because this is normally best reserved for the engineering drawing.

The Menu Facility

Drawing data to a CADD system may be quickly entered (inputted) with the use of a menu. This is a table of drawing commands which may be displayed either on an area of the digitising tablet or directly onto the VDU screen. With a tablet menu (Fig. 9.3), the choice of command is made by contacting the required command square with a digitising pen, whereas a VDU menu requires the use of a keyboard to communicate an alloted command number.

Fig 9.3 Types of menu

(a) COMMAND TABLET MENU

(b) VDU MENU

The individual squares on a command tablet menu are each programmed to carry out a set function and these can be varied depending on the designer's own requirements. For example, a square could be provided to enlarge a part of the drawing, called zooming, in which case the designer just needs to touch the square with the pen and the enlargement occurs.

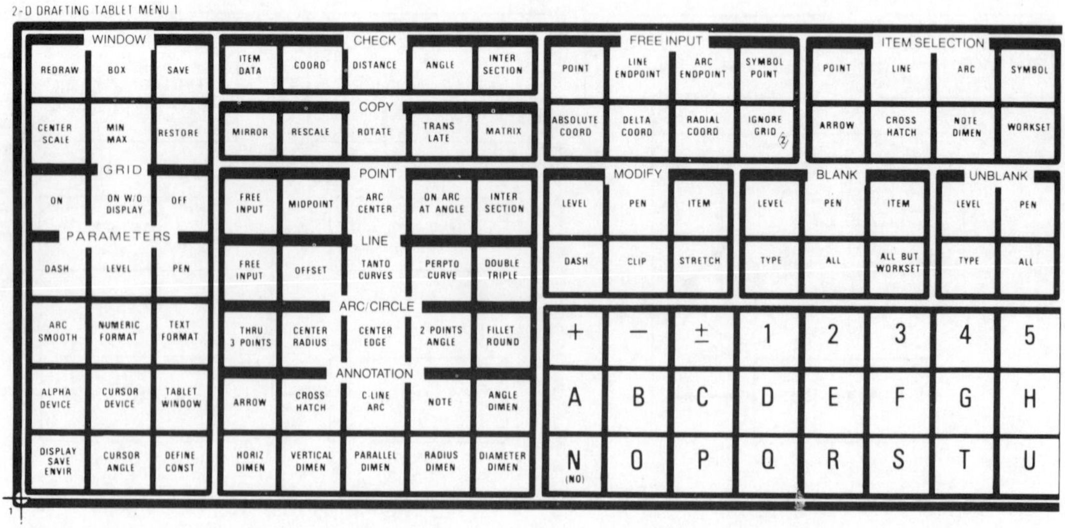

Fig 9.4 Example of a complete command tablet menu board

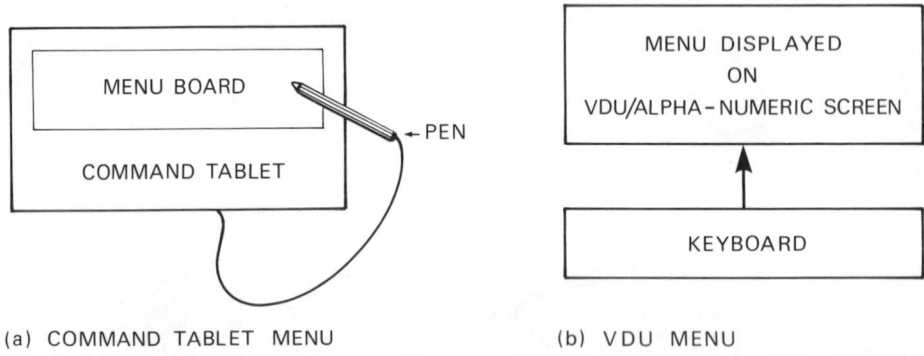

Fig. 9.4 shows a typical command tablet menu board and illustrates the various operations which are possible with it. Designers normally construct their own format to suit the type of work they produce, hence the menu board can include many options.

Positioning and Display of Drawing Elements on the VDU Screen

All drawing matter which may be displayed on the VDU screen is initially devised from four basic elements: points, straight lines, circles and curves.

Points are usually displayed on the screen by selecting the Point command from the menu and stating the required X (horizontal) and Y (vertical) coordinate values of a selected grid position via the keyboard (Fig. 9.5).

Fig 9.5 Grid displayed on VDU screen

Command "Point"
Input "X" Value
Input "Y" Value
Repeat

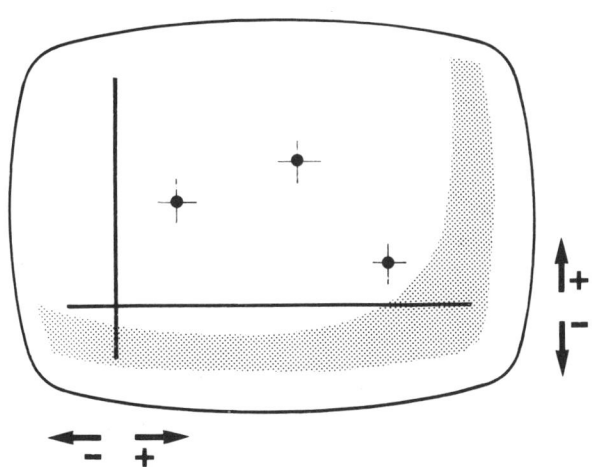

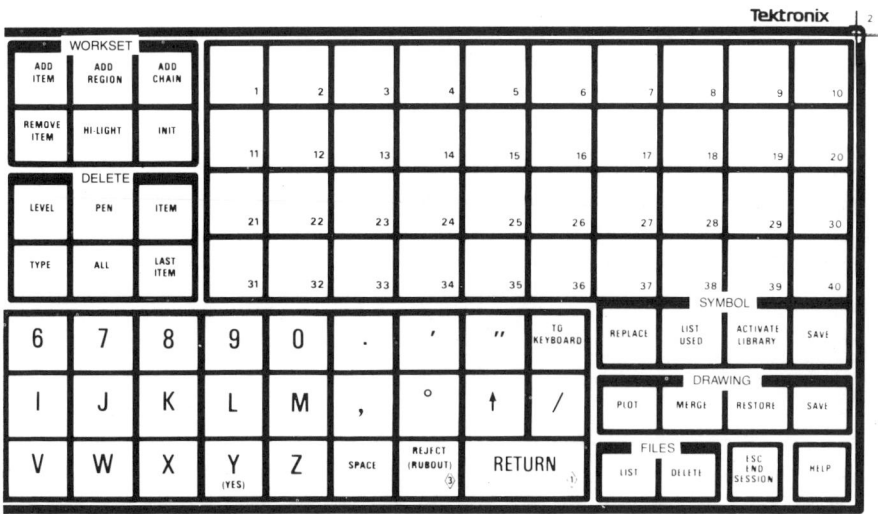

Other drawing elements are then located by moving a pair of cursor cross-hairs around the VDU screen until they are close to the locating point. This may be achieved by

a) Moving a joystick (Fig. 9.6a)

b) Moving a light-pen over the VDU screen (Fig. 9.6b)

c) Moving a pen over an area of the command tablet, separate to that of the menu (see Fig. 9.6c).

Fig 9.6a Location by joystick

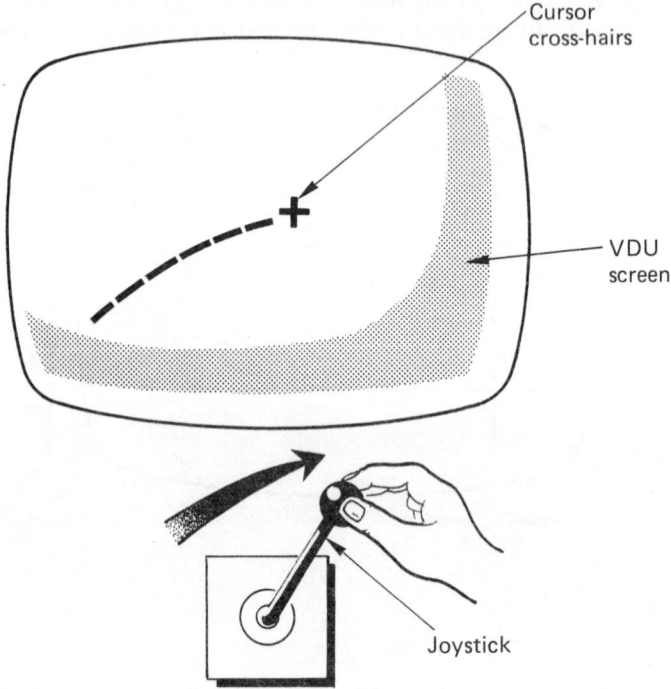

Fig 9.6b Location by light-pen

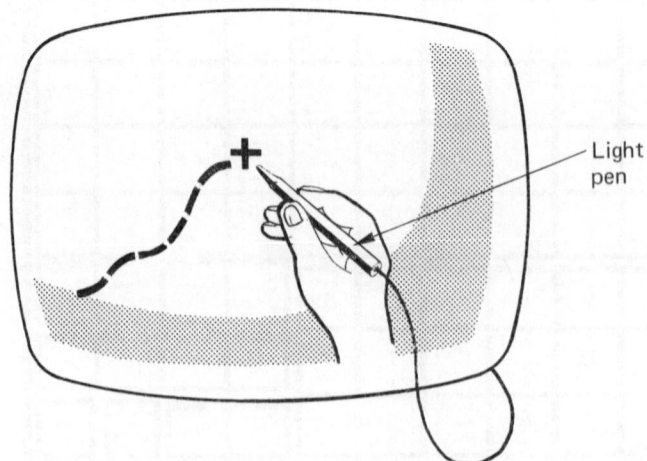

Fig 9.6*c* Location by
command tablet

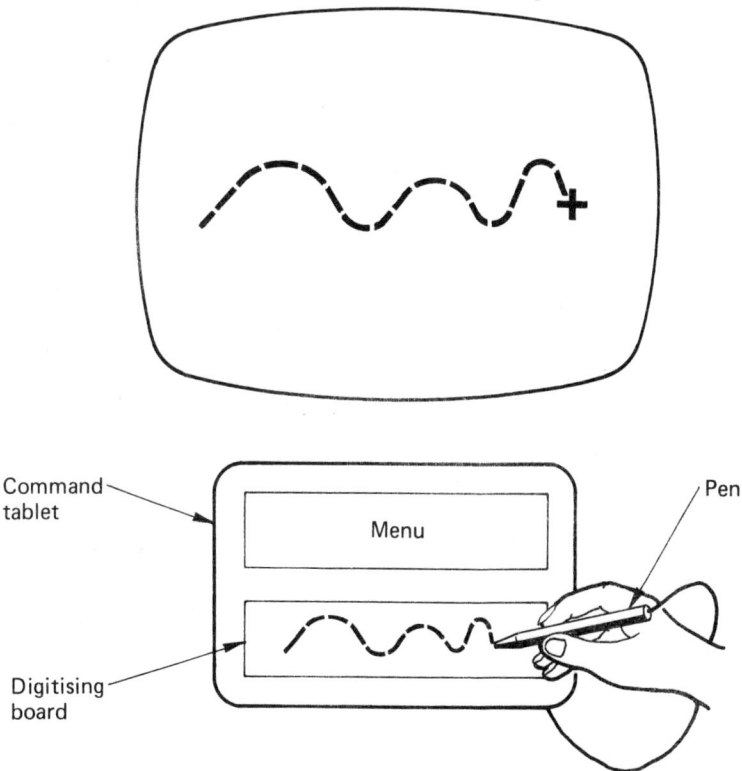

For example, a circle may be constructed by locating a centre point and
quoting a radius. Thus, circles are usually displayed on the VDU screen by
positioning the cursor cross-hairs near the required point, selecting the
Circle command from the menu, and stating the required radius via the
keyboard.

**Shared
Peripherals**

The CADD system is capable of being linked into the manufacture of
machined parts. The design graphics data may be transported from computer,
and translated for feeding into a machine tool controller.

So that the completed engineering drawing can be reproduced for issuing
to the workshop it is necessary to produce a hard copy which is then capable
of being printed. This is done on a plotter. The plotter (or hard copy unit)
can be either a drum type or a flat bed type, and can produce the drawing in
ink or electrostatically. The paper is moved via perforations punched in the
edges of the paper.

Small stepping motors drive the drum in sequence with instructions
received from the computer, and a further stepping motor moves the pen
along the paper and draws the image received from the computer.

There are other shared peripherals which may be connected, such as
plotter/digitiser and line printer, all of which have specific functions to carry
out in the production of the drawing.

Fig 9.6*d* Workstation incorporating light-pen facility

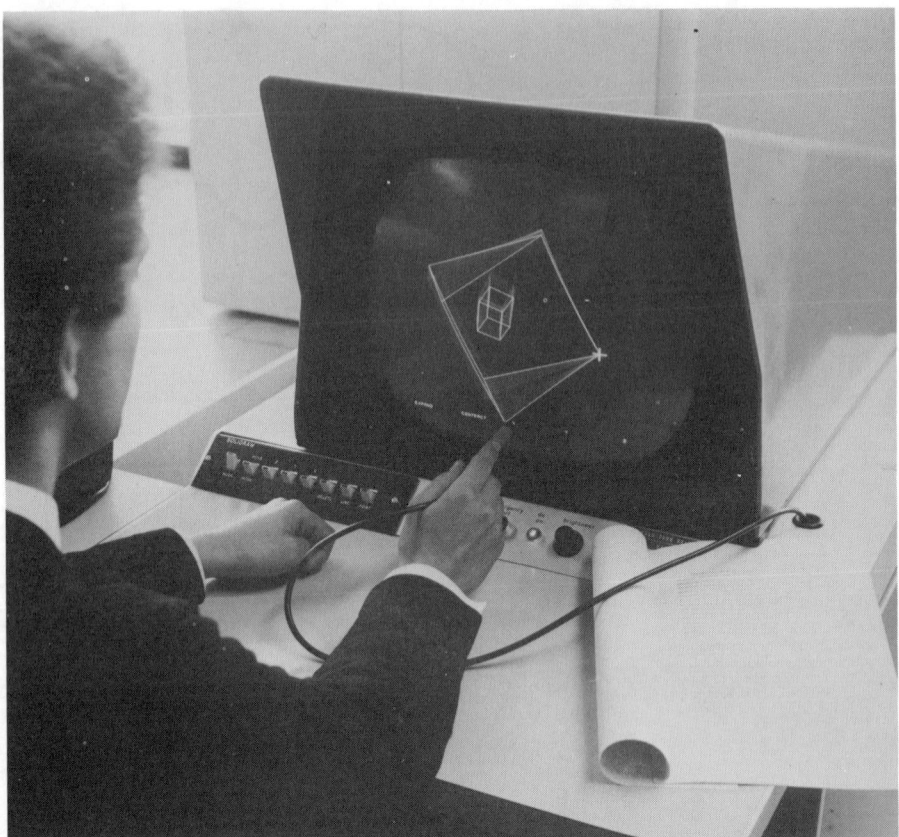

Fig 9.6*e* Workstation incorporating menu board and joystick [*PAFEC*]

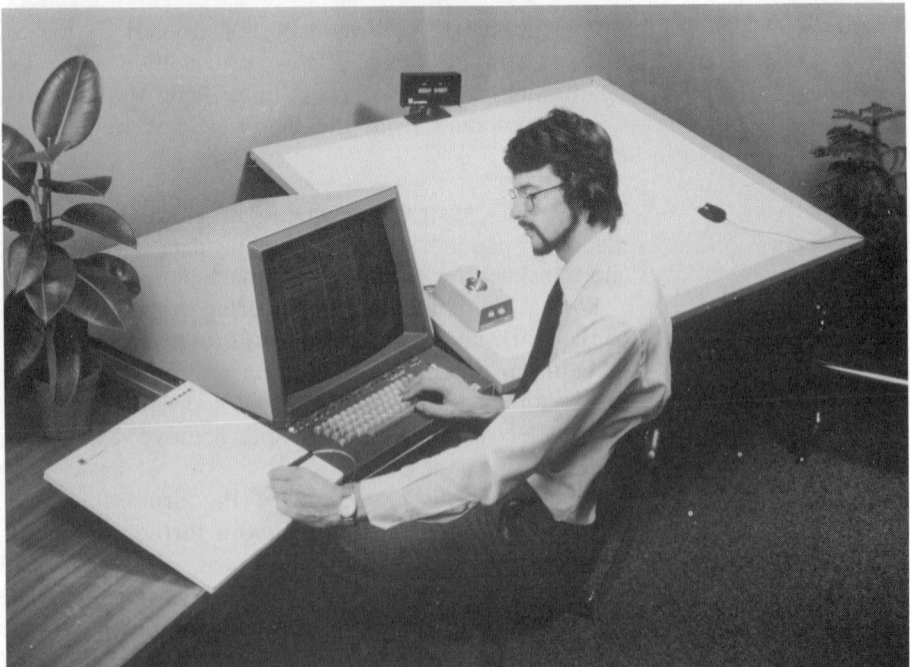

**9.3
Production
and Business
Aspects**

It has always been an essential part of the drawing office to communicate with the shop floor. This was always done through discussions and reference to engineering drawings. From these drawings were produced planning sheets giving details of the method proposed to manufacture the component. The machines were then set by skilled workers to produce the part. It can be envisaged that this process could lead to mistakes and hence incorrect components.

With the advent of the CADD system it is now possible to create machining instructions by the draughtsman producing a drawing on the CADD system, and the resulting graphics data are converted into appropriate manufacturing codes. These codes can then be loaded into "computer-aided part programming" software, where manufacturing information, such as tool selection and feed rates may be added. This completes a "part program" for producing an engineering component, whose coded data may be transported directly by cable link, from computer to a numerically controlled (NC) machine tool. The machine tool is then capable of converting the coded instructions into machining operations and the component is produced by the robotic action of the tools.

Also, it is now possible to create machining instructions by the draughtsman producing a drawing on the CADD system, and coded information being fed directly to a tape machine where punched tape is produced ready for inserting into the NC machine tool. The job of the production engineer who once produced the planning sheets is now largely superseded by a design engineer capable of understanding the relationship between drawing requirements and the machining process.

With the highly competitive marketplace that exists in the world to-day, engineering companies wishing to sell their products need to complete their specifications and designs fast enough to compete with their rivals on price and delivery dates, etc. By using a CADD system, rapid tendering on projects is possible and it allows the engineers the facility to look at many options and hence come up with the ideal solution regarding feasibility, design excellence and cost.

**9.4
Computer-aided
Engineering**

Computer-aided engineering is the term used to describe the integration of all departments within a manufacturing company in so far as they are all linked to one computer. Each department can input and output its own data and can also integrate the information input by other departments.

The main departments within an engineering company are engineering design, production engineering, sales, marketing, purchasing, quality, despatch and accounts. Therefore, a design can be completed and it is then possible for production engineering to retrieve information from the computer and use it for deriving their manufacturing data. Similarly, the purchasing department can retrieve a list of parts contained on the design and use this to place orders for parts. The sales department can establish the product specification and function characteristics, and they are also able to read information input by the accounts department.

It is therefore a very powerful system and must be managed very carefully, otherwise it is possible to input and output the incorrect information and cause chaos in the company. Fig. 9.7 shows a diagrammatic representation of the system.

Fig 9.7
Computer-aided
Engineering:
representation of
system

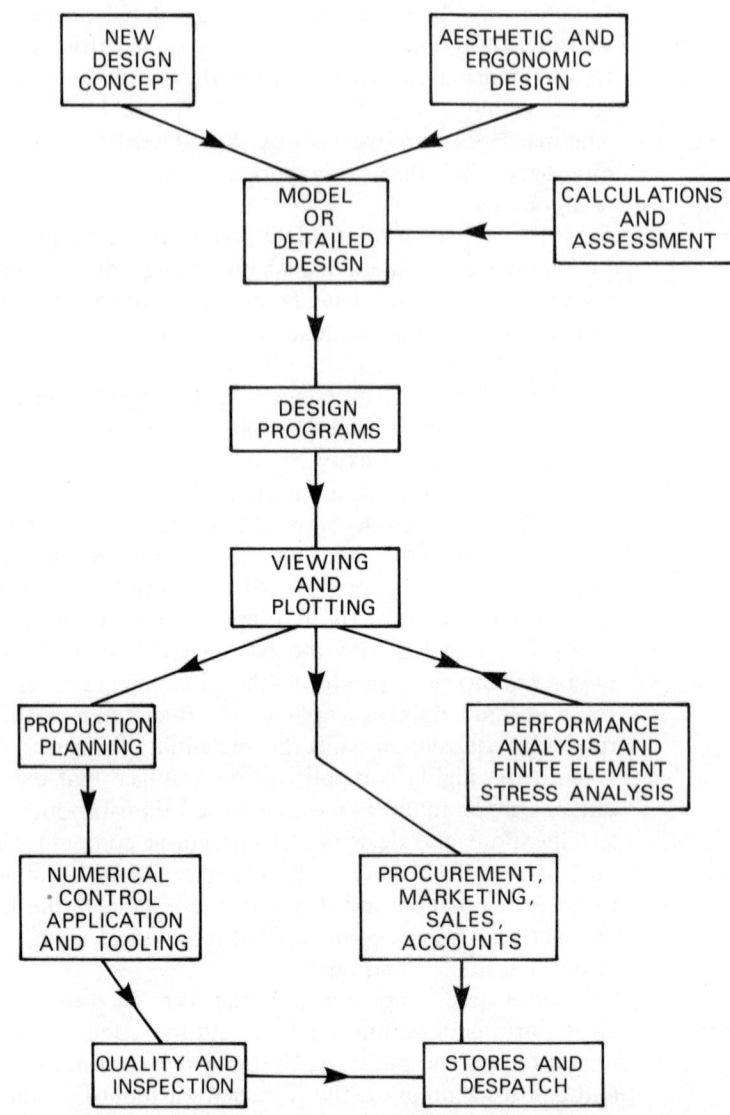

**9.5
Applications of
CADD**

Some specific practical applications and techniques of CADD will now be discussed.

1 *2D draughting techniques* As already outlined, 2D geometric shapes may be developed by using combinations of the basic drawing elements. Also many CADD systems employ various additional draughting techniques.

2 *Automatic fillet radii* may be displayed by locating the required corner with the cursor cross-hairs and inputting the radius size (Fig. 9.8). Usually the corners are automatically erased as part of the operation.

Fig 9.8 Fillet radii

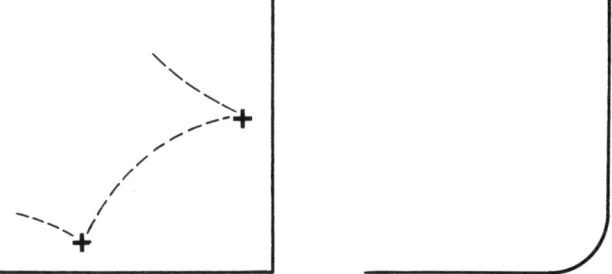

3 *Automatic 45° chamfers* may be displayed in a similar fashion to fillet radii (Fig. 9.9).

Fig 9.9 45° chamfer

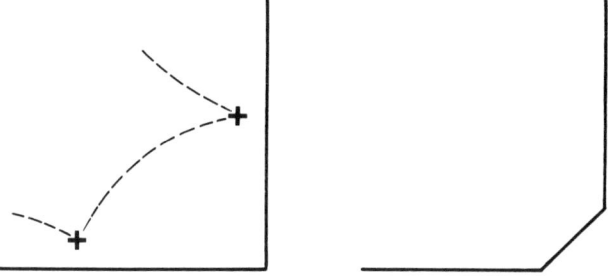

4 *Automatic cross-hatching* is a time-saving facility. The area to be hatched is located with the cursor. Most systems will then automatically scan the perimeters of the indicated area and hatch within those perimeters (Fig. 9.10).

Fig 9.10
Cross-hatching

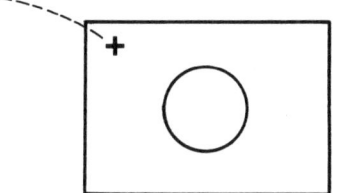 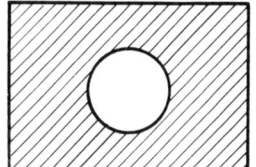

5 *Automatic dimensioning* may be displayed by locating the required edges and the position of dimensions. Most systems then automatically calculate the size and draw the dimensions, leaders and arrowheads (Fig. 9.11).

Fig 9.11
Dimensioning

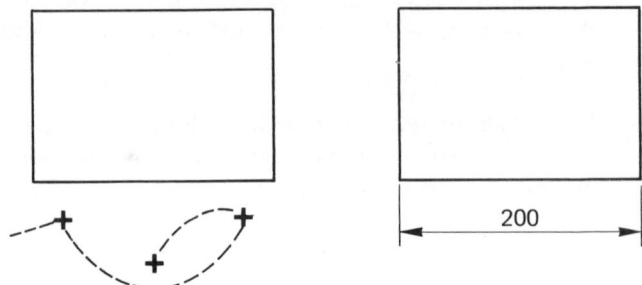

6 *Editing* Lines, or portions of the drawing, may be erased at any stage in its development so that modifications can be made. Dimensions and cross-hatched areas are also easily erased and modified.

7 *Zooming and panning* (Fig. 9.12) Individual localities of the drawing may be enlarged (or "zoomed") to any required scale. This is usually available for observing or editing intricate details. A zoomed area may be panned in a required direction to observe or edit other features.

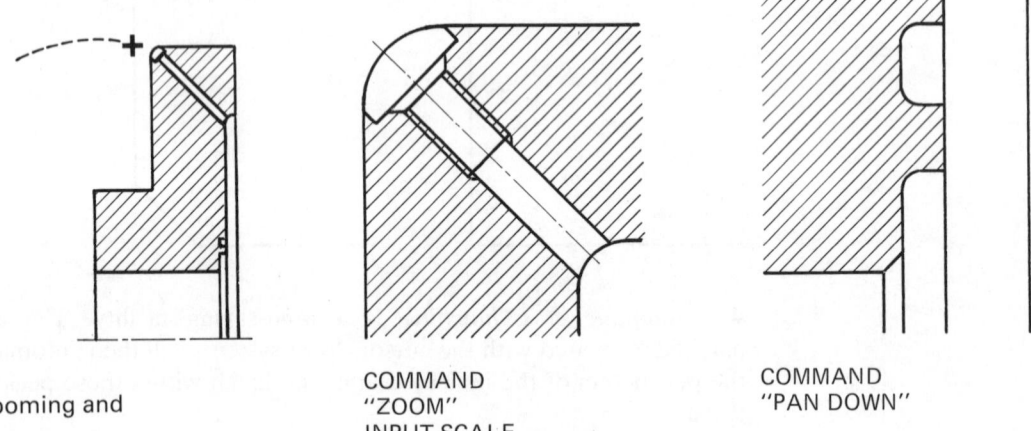

Fig 9.12 Zooming and panning

COMMAND
"ZOOM"
INPUT SCALE

COMMAND
"PAN DOWN"

8 *Transformations* Apart from zooming and panning, many systems incorporate other transformation facilities such as rotation about a point, mirror image, and transformation repetition. Fig. 9.13 shows how a drawing may be given a mirror image and then the operation quickly repeated to complete most of a tooth-rack profile.

Fig 9.13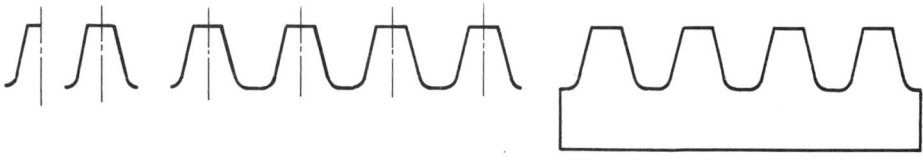

START MIRROR   REPETITIONS   COMPLETE
IMAGE

9 *Layering* It is often possible to display drawings with several component parts, as a series of layers which may be viewed separately or in
combination. This technique is commonly employed for assembly drawings
with varying cross-sections, and in printed circuit board (PCB) applications.
Fig. 9.14 shows two layers of a temperature controller display panel.

Fig 9.14 Temperature
controller display
panels [*Computer
Aided Design Centre,
Cambridge*]

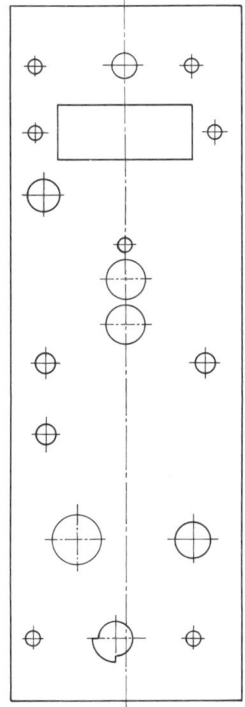
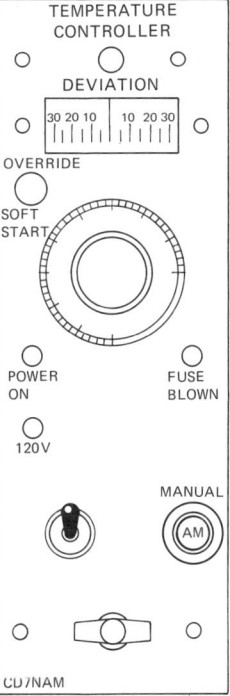

10 *Symbols and standards libraries* Libraries of British Standard symbols
such as BS 308, Electrical, Welding, Hydraulic, etc. may be stored in the
computer memory and individual symbols quickly recalled to the VDU
screen when required.

Also, libraries of standards components, such as nuts and bolts, bearings,
etc., may be compiled. Any drawing, or part of drawing, may be added to a
standards library to eliminate repetitive drawing work. Fig. 9.15 shows a
typical example.

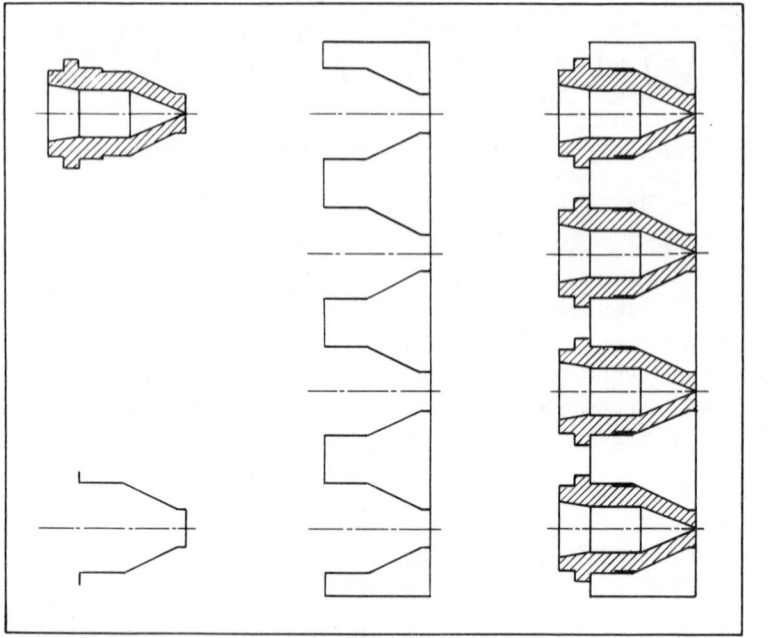

Fig 9.15 Part of standards components library [*Computer Aided Design Centre, Cambridge*]

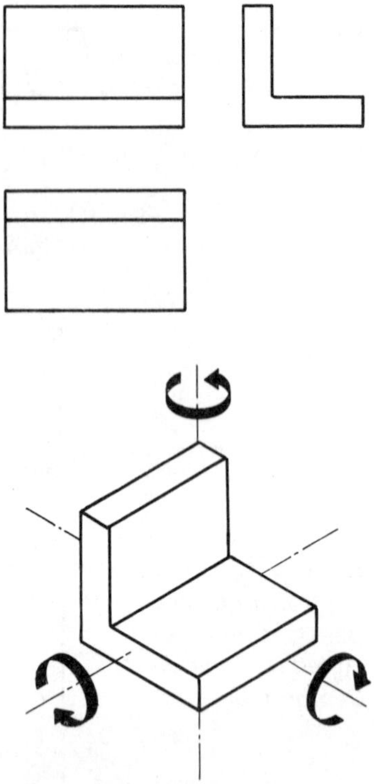

Fig 9.16 3D display of 2D orthographic views

11 *Parts lists and text libraries* By storing parts lists and materials stock lists, in the computer memory, permanency and savings in time and space are readily achieved. The required lists may be quickly recalled to the alpha-numerics screen. Also, libraries of standard text may be compiled for quick recall to the VDU screen.

12 *Basic design calculations* Software for common design calculations may be compiled or purchased to give another time-saving aspect of CADD. Typical applications could include calculations for volume, mass, centroid positions, second moments of area, beam reactions, simple bending stresses, gear loads and bearing lifetimes.

13 *3D draughting and design* A 3D CADD system is one which can recognise a number of 2D orthographic views in 3D form. It should then be possible to display a 3D form on the VDU screen for given 2D views (or vice versa) (Fig. 9.16). Some systems also have the facility to rotate the 3D view about any of its three axes, and to convert orthographic assemblies into exploded 3D views.

14 *Geometric modelling*   More complex shapes may be visualised at the design stage by 3D geometric modelling techniques. The most common types may be listed as

a) 3D wireframe models

b) 3D solid models.

Wireframe models enable the designer to create intricate profiles by locating various points in three planes, and plotting "wire" profile lines through them. Once displayed, the model may be sectioned, intersected, blended with other shapes, and viewed in any direction. Fig. 9.17 shows typical examples of the wire mesh models which can be created for mouldings, castings and pressings.

Fig 9.17 Wire mesh models [*Delta Computer Aided Engineering*]

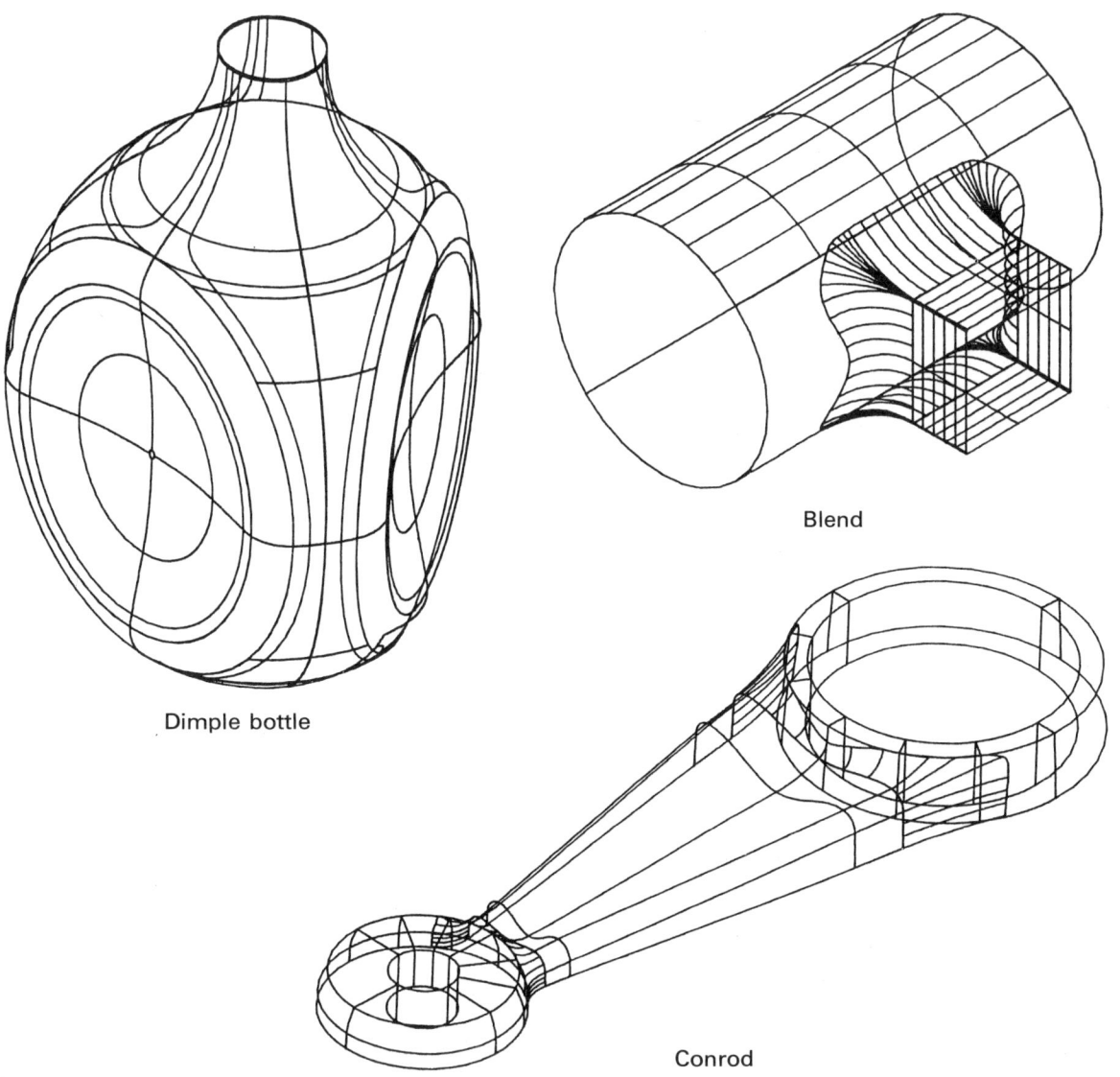

Dimple bottle

Blend

Conrod

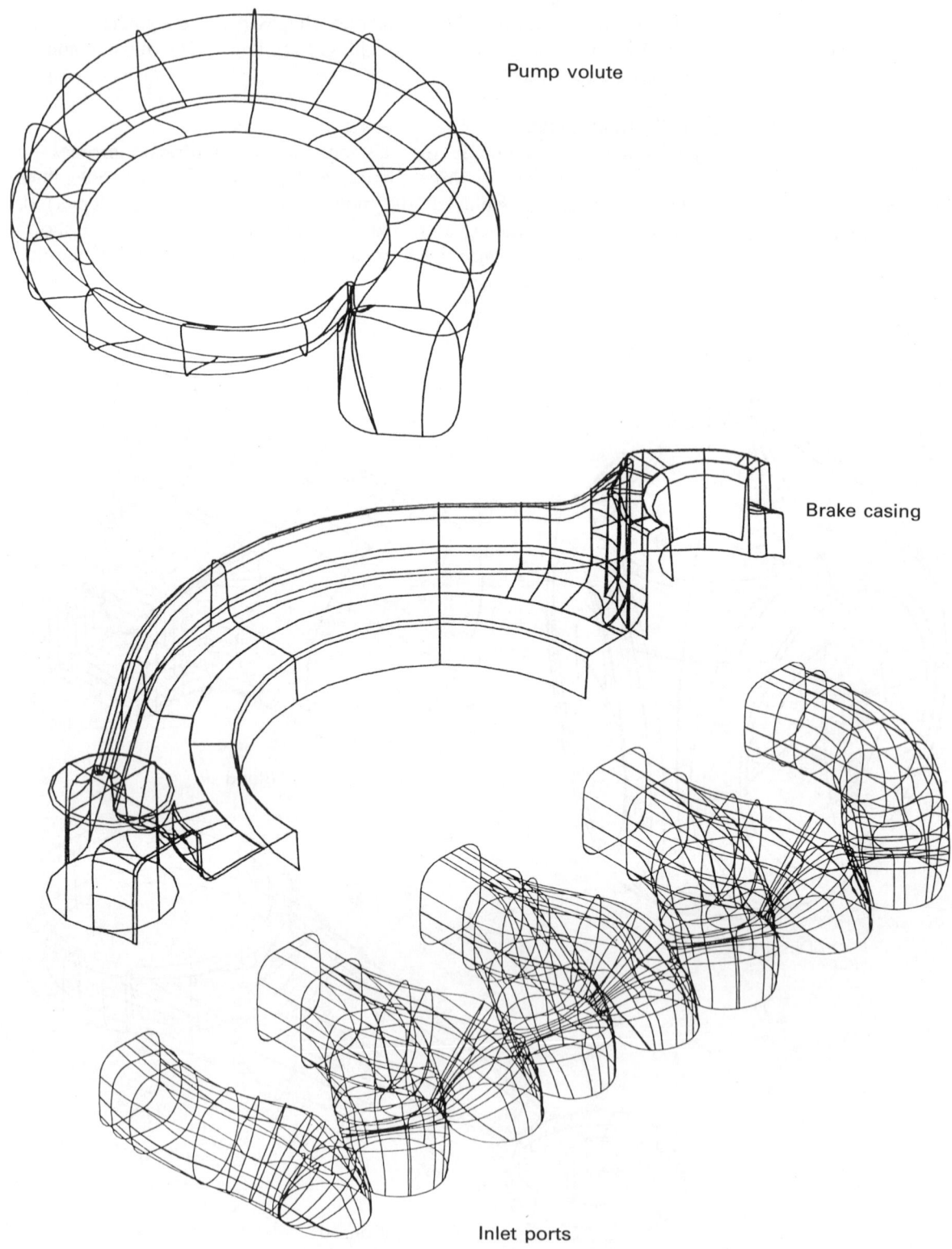

Pump volute

Brake casing

Inlet ports

Despite the use and versatility of wireframe models they can be limited by a certain amount of ambiguity in shape and lack of surface information. When using solid models, a complete geometric description of a part is generated. Systems with a solid model facility will usually incorporate extensive colour choice and tone control to give improved visualisation of shape and ready identification of components, surfaces and cross-sections. A solid model also has analytical advantages—particularly in mass property calculations and the construction of finite elements. Fig. 9.18 shows a good application of solid geometric modelling.

Fig 9.18 Solid geometric modelling [*Computervision Corporation*]

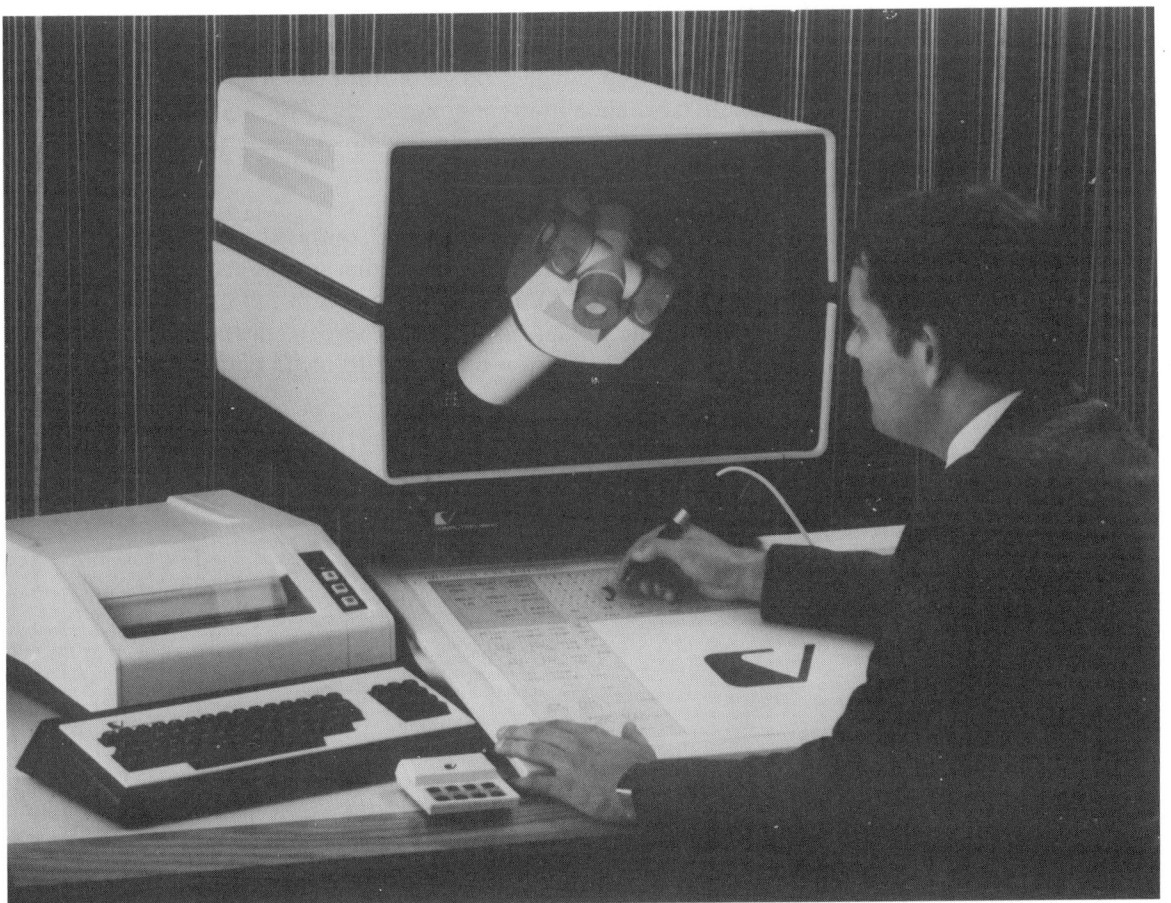

15 *Finite element analysis* This is an appropriate method of analysing a wide variety of mechanical properties of complex components and structures. Analysis of stress patterns and deflection characteristics are typical applications.

The basic principle of finite element analysis is to split a complex shape into a number of smaller, simpler elements. Typical element shapes are shown in Fig. 9.19. These finite elements are then analysed for their strain characteristics and the results are related back to the whole structure. Fig.

9.20*a* shows a 3D stress analysis using triangular elements, on a more advanced procedure called "boundary element mesh".

The calculations involved with this type of analysis would be extremely time-consuming without the use of a CADD system.

Fig. 9.21 shows a 2D stress and deflection analysis of a hook.

16 *Ergonomic applications* Chapter 6 outlined the many uses of anthropometric data, and 2D cardboard manikin models, but also pointed out their limitations. The infinite variety of human limb sizes and arcs of movement can be considered far more efficiently by using appropriate CADD software. Figs. 9.22 and 9.23 are extracts from a software package called "Sammie" which shows how anthropometric data may be stored in the CADD system to produce a range of intricate 3D "man-models" which give invaluable assistance in ergonomic design. [These diagrams are shown by kind permission of Compenda Ltd., who produce the "Sammie Package".]

17 *Plant and pipework layout* Chapter 7 outlined the importance of 3D layout models in these applications, particularly when components need to be re-arranged or deleted. A CADD system with 3D graphics enables the designer to create the model on the VDU screen. In the case of pipework layouts, the model is then often converted to a 3D detail drawing.

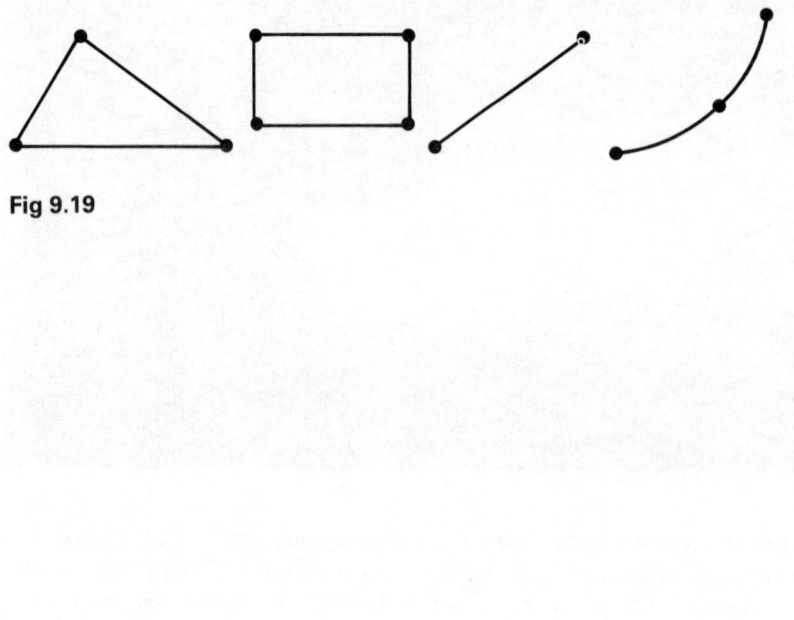

Fig 9.19

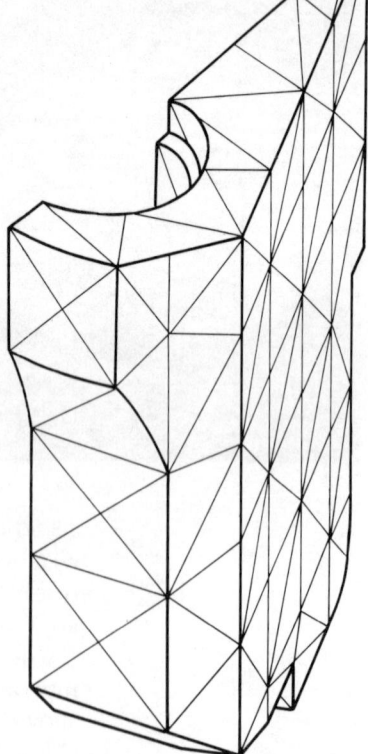

Fig 9.20*a* 3D stress
analysis
[*Computational
Mechanics Centre*]

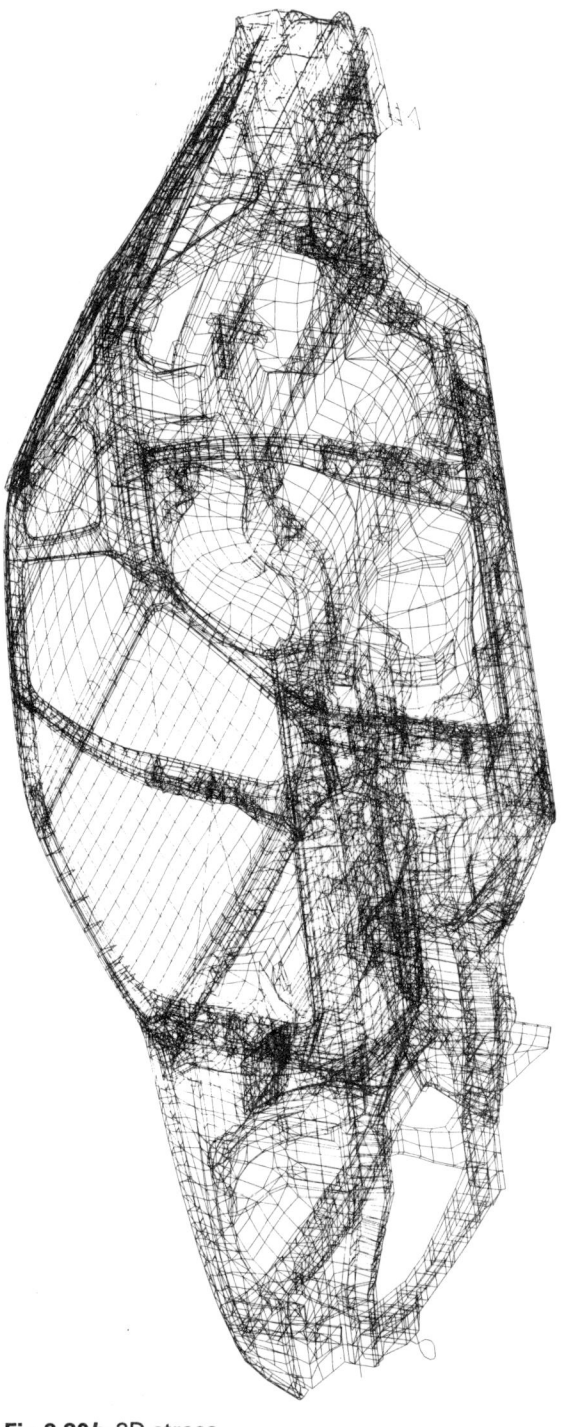

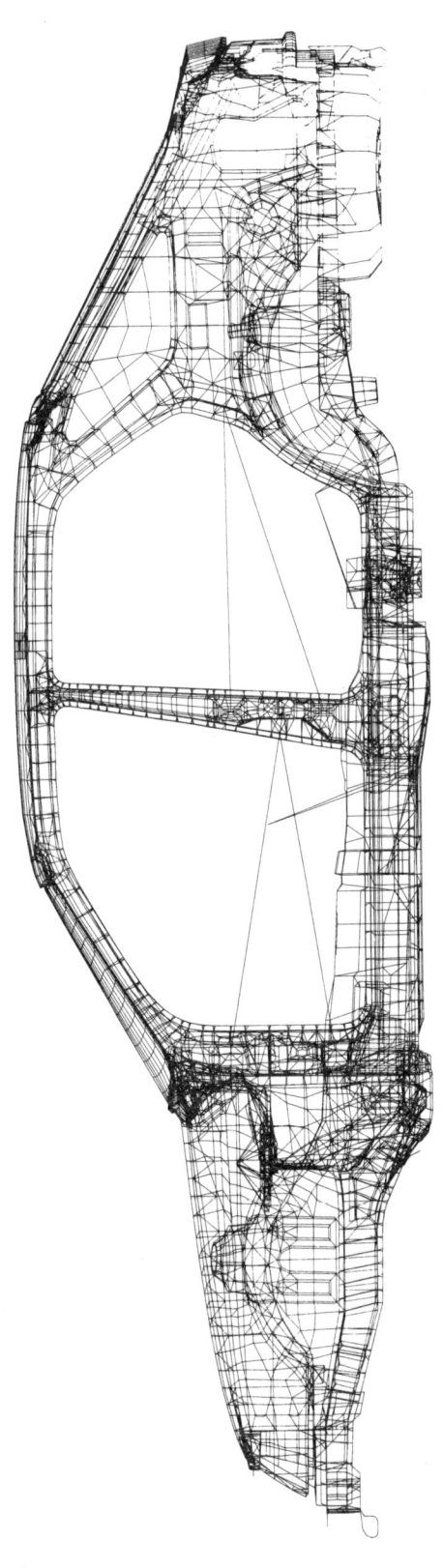

Fig 9.20*b* 3D stress analysis for the Ford Sierra [Ford Motor Company]

Fig 9.21 2D stress analysis [*ECS*]

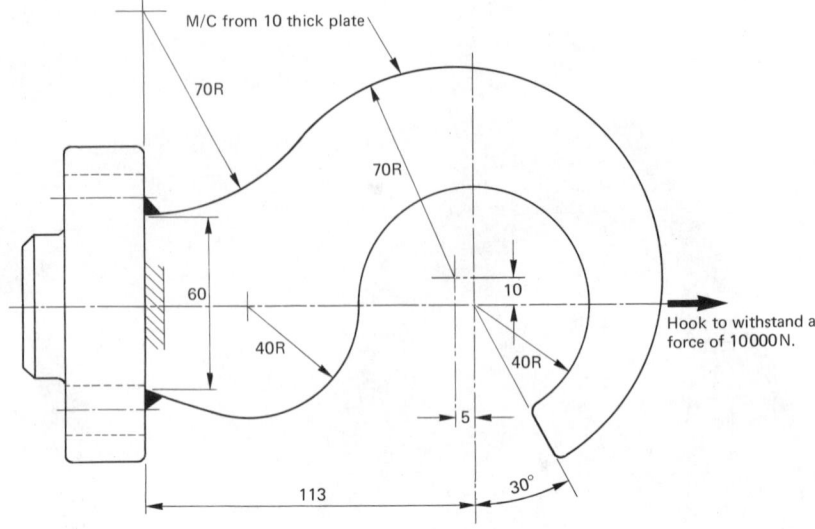

STEP 1

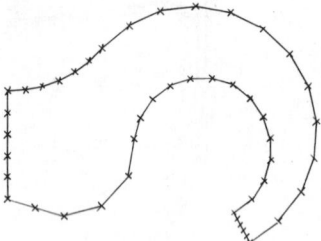

Finite Element Boundary

STEP 2

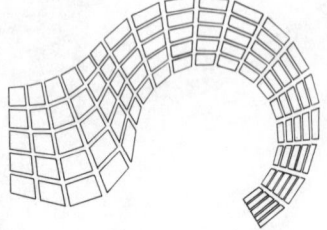

Finite Element Mesh

STEP 3

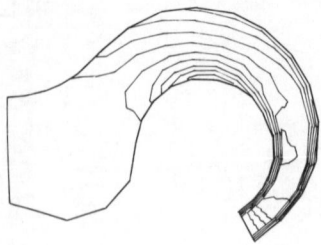

Stress Contours

STEP 4

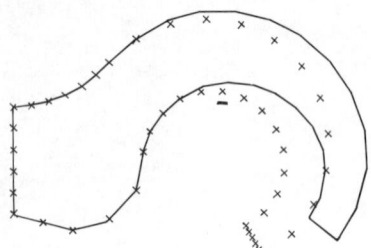

Deflect Shape

Fig 9.22
Anthropometric
data [*Compenda*]

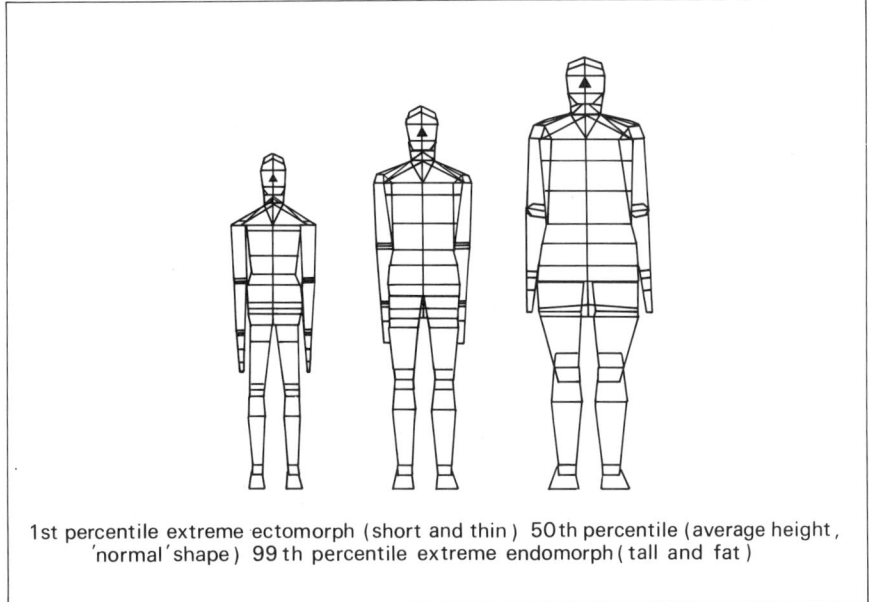

1st percentile extreme ectomorph (short and thin) 50th percentile (average height,
'normal' shape) 99th percentile extreme endomorph (tall and fat)

Fig 9.23 Simulated
cockpit situation, a
confined working
space [*Compenda*]

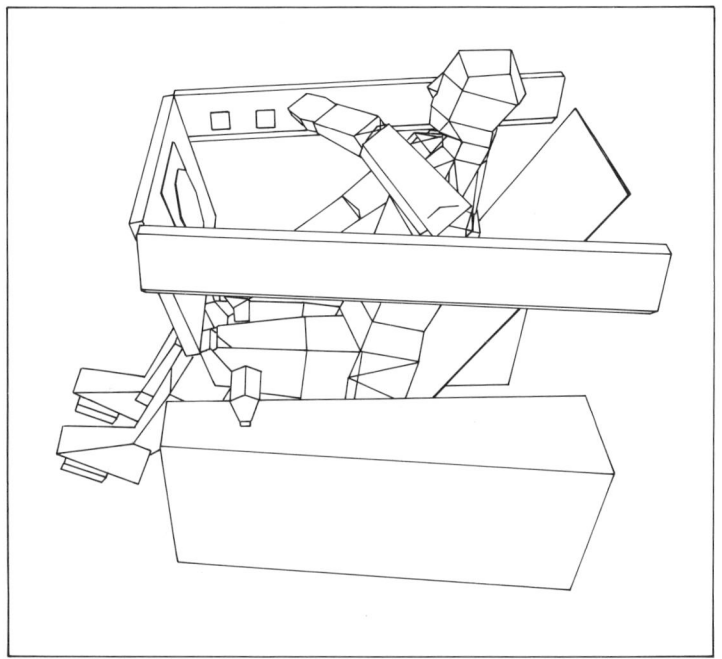

10 The Complete Design Process—a Case Study

This chapter inter-relates some of the previous discussions into the description of a complete design project.

DIY CONCRETE MIXER

Benford Ltd. of Warwick produces a wide range of construction equipment and currently employs upwards of 700 people. This company has recently made a successful bid to capture a share in the market for a comparatively new design concept—a low-priced concrete mixer aimed specifically at DIY use as opposed to the building site.

Close to the start of the project, a directors' board meeting, which included invited guests, took place.

BOARD MEETING
Present   Managing Director
          Sales Director
          Technical Manager
          Works Director
          Market Research Manager
The outcome of this meeting was
   a) A decision to design the mixer.
   b) The drawing-up of a specification for it.
This specification is to be based on recommendations in the British Standards publication PD6112 *Guide to the Preparation of Specifications*, as mentioned in Chapter 1.

**10.1
Product
Specification**

Title   LOW-PRICE "LITTLE BENFORD" TIP-UP CONCRETE MIXER (3/2 SIZE)

List of contents
   1 History and background information
   2 Scope of specification
   3 Conditions of use
   4 Characteristics
   5 Ergonomic considerations
   6 Appearance (Aesthetics)
   7 Performance
   8 Reliability and life
   9 Packaging
   10 After-sales service.

Fig 10.1 Product range [*Benford*]

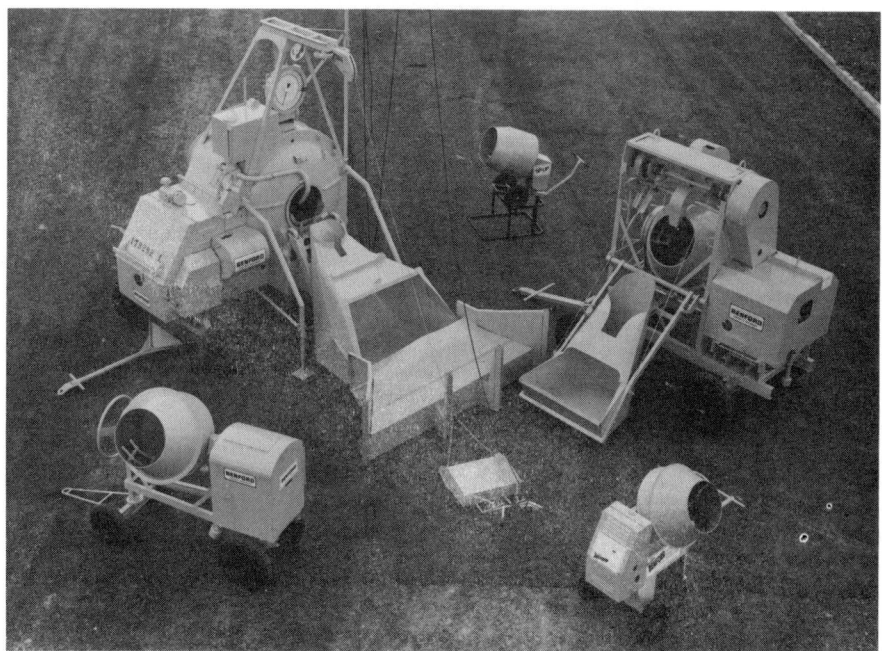

1   *History and Background Information*

Our recent market research department survey has indicated a remaining gap in the market for low-capacity concrete mixers suitable for the practical do-it-yourself enthusiast.

This gap is for a tip-up model of a price low enough to attract full purchase by the user, whilst also providing a "smaller brother" of our Half-bag Tip-up model which is sold mainly to plant hire firms.

The suggested price is in the region of £160 per unit for a sales forecast of 5000 units per year.

It is further recommended that the mixer be compact, light, and easily dismantled for transporting in a small estate car.

2   *Scope of Specification*

The final specification cannot be concluded until

  a) A prototype model has been designed, manufactured, and tested.
  b) Cost analysis is completed, and customer feedback obtained after a
      trial batch of two or three units has been sold.

However, the basic requirements are as follows:

Capacity   60 litres mixed batch (an average barrow load)
Dimensions   420 mm diameter drum opening.
              1050 mm total height.
              610 mm maximum width at wheel axle.
Drum speed   approximately 30 rev/min.
Assembly   All items easily assembled and dismantled by the average
              DIY person using a tool kit provided.
Drive   0.25 kW, single phase, 50 Hz, electric motor speed
          1450 rev/min (suitable for domestic supply).

3 *Conditions of Use*

The product must be able to mix quantities of cement, water, sand aggregates, or plaster in varying proportions, under its own power, without spillage.

The mouth must be easily loaded with a shovel, and enable fast emptying without leaving the drum clogged.

The complete unit must be light and portable enough for easy manoeuvrability around a garden or work area. It must also be easily transported in the back of a mini estate car with the tilt-handle removed.

With future export sales in mind, the local climate should have no adverse effect on performance and reliability.

4 *Characteristics*

Mixing Drum To be either injection-moulded plastic base and cone or welded spun steel base and cone, and two heavy-duty mixing blades bolted base and cone. Drum to be supported on plain bearings if possible.

Drive Speed reduction from electric motor to be provided by gears or a combination of gears and vee-belts.

Frame Welded steel fabrication from angle, plate, tube, or a combination of each. Structure to be of minimum material, whilst maintaining the strength and stiffness for the conditions of use.

Tilt-handle Detachable from mainframe for ease of car transport. Made from circular steel tube for ergonomic and aesthetic reasons, to be long enough for adequate tipping leverage by a 5th percentile man, and small enough to accommodate a human hand.

Wheels Two wheels of diameter 200 mm, made from plastic injection moulding or welded steel plate fabrication.

Guard Welded steel plate fabrication.

Tool kit Mixer is to be supplied with a set of tools and simple assembly instructions.

5 *Ergonomic Considerations*

These are specified elsewhere but may be summarized thus:

    a) Size and height of drum for filling.
    b) Ease of manoeuvrability.
    c) Ease of tipping.
    d) Comfortable grip of tilt-handle.
    e) Ease of assembly and maintenance.
    f) Size and comfort of tools.
    g) Weight of dismantled parts.

6 *Appearance* (*Aesthetics*)

The product should look symmetrical and stable and give the impression of precision and efficiency of function, with pleasing continuity of profile.

Aimed directly at the domestic market, it should have a refreshing up-dated sense of styling, as suitable for display in a catalogue as any power drill, work bench, foodmixer, or vacuum cleaner.

Flat surfaces and smooth lines will aid cleaning.

Steel surfaces will be given a double coating of prime and paint for protection against weather and possible rough treatment. Inside of drum need only be primed. Pleasing colour contrast will be obtained with black frame and wheels, and our standard bright yellow on drum and guards.

7 *Performance*

The prototype mixer will be tested for performance under similar conditions to those expected in service.

Should generally cope with overloading, but for serious cases, the motor is fitted with thermal overload and reset button.

8 *Life and Reliability*

The mixer would be expected to achieve a useful life of 7 years under full loading capacity for a quoted number of hours per day and days per year.

To this end, the prototype model will be tested under full load and intermittent stopping and starting for a quoted number of hours.

During the test period, lubrication and the changing of spare parts should conform to maximum frequencies which will be quoted in the final specification.

9 *Packaging*

The mixer will be supplied to the customer in totally dismantled form and packaged in a cardboard carton which is compact and easily handled.

10 *After-sales Service*

All servicing should be easily done by the customer with the tool kit and instruction booklet provided.

The following items will be available as spare parts:

Belts and Pulleys (if used)

Pinion Shafts and Drum Shafts

Motor

Bearings.

In the event of accidental damage, the following items should be available in smaller quantities:

Drum

Wheels

Guard

Tilt-handle.

10.2
Planning the
Project

The Technical Manager now prepared a planning schedule to determine a design time for completing a fully-tested prototype mixer. The alloted tasks were listed thus:

1. Order motor ($\frac{1}{2}$ week)
2. Motor delivery (4 weeks)
3. Drive synthesis and evaluation ($\frac{1}{2}$ week)
4. Order belts and pulleys (if used) ($\frac{1}{2}$ week)
5. Delivery of belts and pulleys (2 weeks)
6. Drum cost analysis ($\frac{1}{2}$ week)
7. Order drum ($\frac{1}{2}$ week)
8. Drum delivery (6 weeks)
9. Wheels cost analysis ($\frac{1}{2}$ week)
10. Order wheels ($\frac{1}{2}$ week)
11. Delivery of wheels (5 weeks)
12. Select and order bearings ($\frac{1}{2}$ week)
13. Bearings delivery (3 weeks)
14. Frame and handle synthesis ($\frac{1}{2}$ week)
15. Frame comparison and evaluation (1 week)
16. Design calculations (1 week)
17. Design drawing (1 week)
18. Overall cost analysis (1 week)
19. Re-design for cost (if necessary) ($\frac{1}{2}$ week)
20. Detail drawings (1 week)
21. Make frame, handle, gears, shafts, guards, tools (2 weeks)
22. Assemble ($\frac{1}{2}$ week)
23. Test (3 weeks)
24. Re-design for function (if necessary) (1 week)

A network analysis chart was then drawn up as shown in Fig. 10.2 and based along the lines described in Chapter 2.

This indicated a prototype design time of $12\frac{1}{2}$ weeks with the critical path lying along tasks 14, 15, 16, 17, 18, 19, 20, 21, 22, 23, and 24.

10.3
Delegation of
Design Work

The Technical Manager has overall responsibility for the drawing office at Benfords, which employs a combination of design and detail draughtsmen.

For this project one senior draughtsman was delegated with most of the initial design investigation, whilst another draughtsman was available to help with detail work at a later date.

Synthesis of
Frame and
Tilt-handle

Synthesis involves collecting together design information and ideas. Fig. 10.4 shows some of the options considered for the tilting handle (all from circular steel tube as specified). Likewise, Fig. 10.5 shows some ideas for the design of the mainframe.

At this stage, the main object was to compare stiffness of structure (torsion and bending) with quantity of material (the main indication of cost here). To achieve this, some preliminary shapes similar to those shown were quickly fabricated in the company's Research and Development Depart-

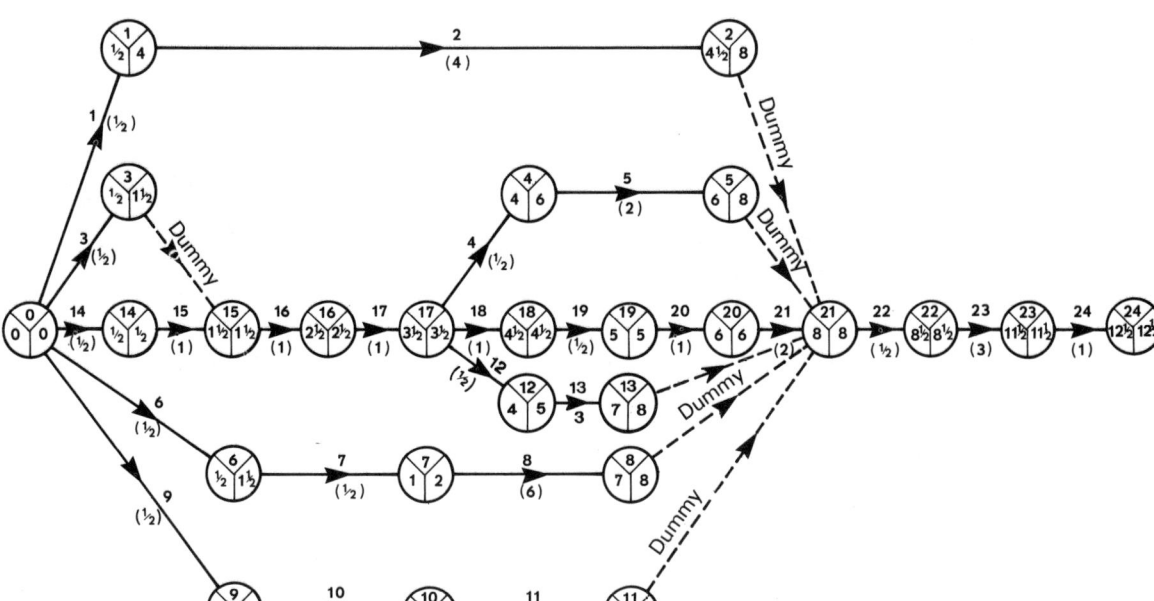

Fig 10.2 The network
analysis chart

Fig 10.3 The
draughting stage

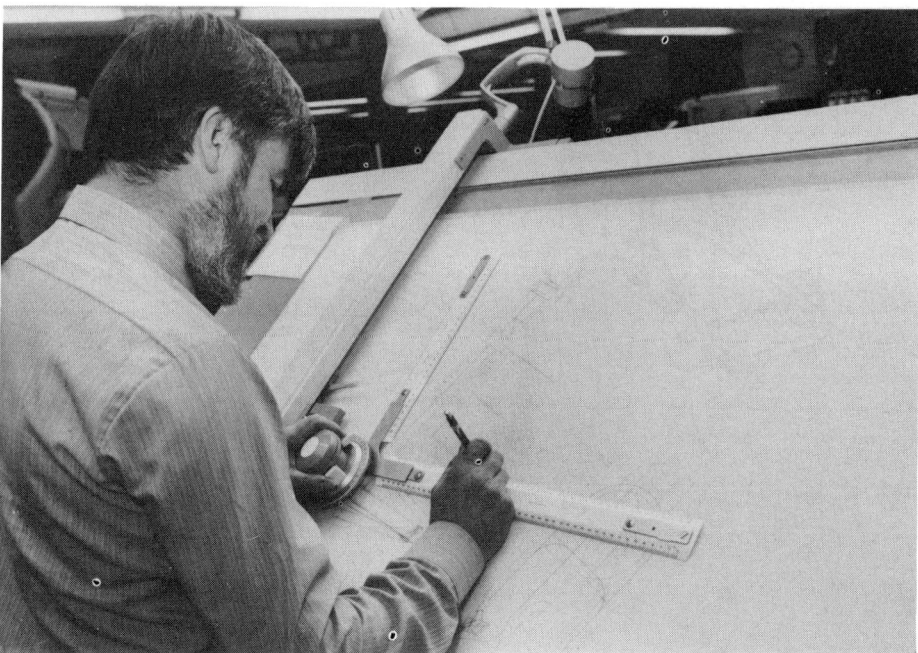

Fig 10.4

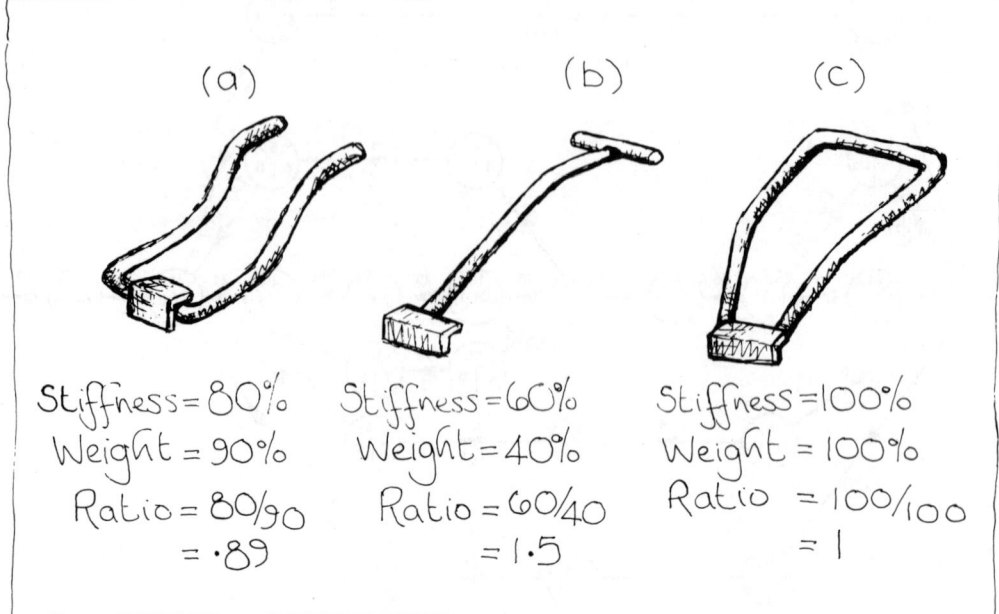

	(a)	(b)	(c)
	Stiffness = 80%	Stiffness = 60%	Stiffness = 100%
	Weight = 90%	Weight = 40%	Weight = 100%
	Ratio = 80/90	Ratio = 60/40	Ratio = 100/100
	= .89	= 1.5	= 1

Fig 10.5

(a)

Circular Tubes with Gussett Plates

Stiffness = 100%
Weight = 100%
Ratio = 100/100
= 1

(b)

Circular Tube with Folded-Plate-Pedestal

Stiffness = 80%
Weight = 85%
Ratio = 80/85
= .94

(c)

Square Tube Single-Stem Structure

Stiffness = 75%
Weight = 65%
Ratio = 75/65
= 1.15

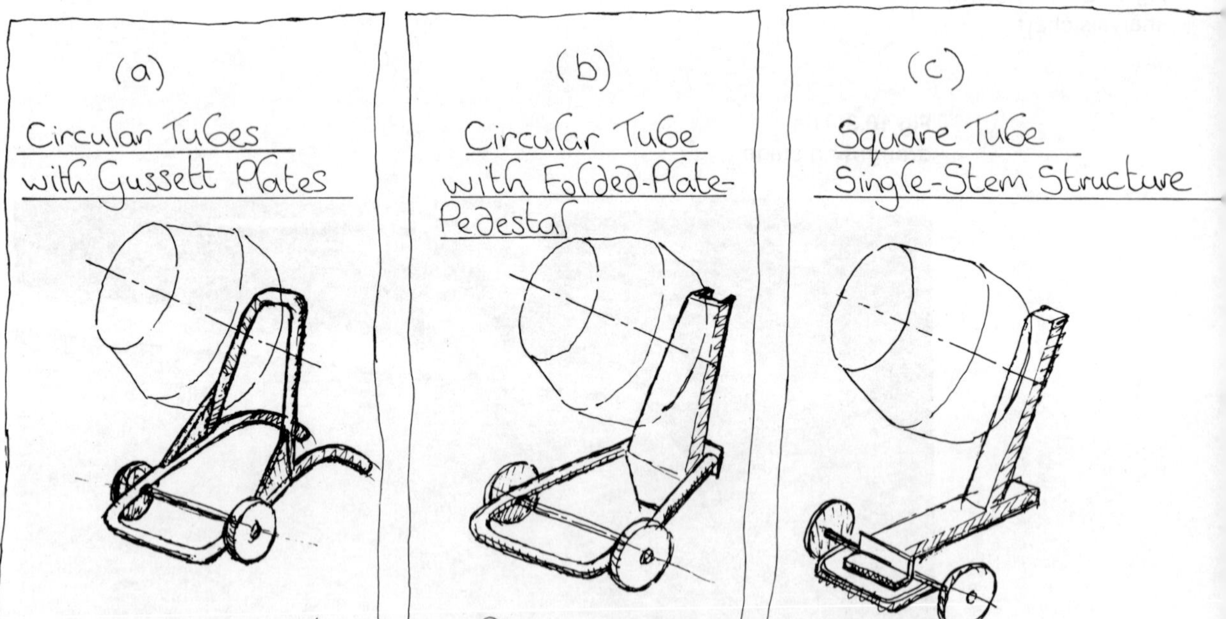

ment, working from the rough sketches and under close liaison with the designer. Each shape of handle and frame was made to the same outside dimensions and similar cross-sectional size of elements. Maximum torsional and bending deflections were obtained under identical loading of individual shapes and combinations of handle and frame.

Comparison and Evaluation of Frame and Tilt-Handle

The comparative results of stiffness/weight ratio in Figs. 10.4 and 10.5 show idea (b) to be the most favourable handle design and idea (c) the most favourable frame design. These two solutions also gave the best results in combination.

Other aspects, such as Safety, Ergonomics, and Appearance were also considered in a systematic scheme similar to that shown in Chapter 1 The chosen solution was endorsed as a combination of idea (b) for the handle and idea (c) for the frame, as shown in Fig. 10.6.

Fig 10.6

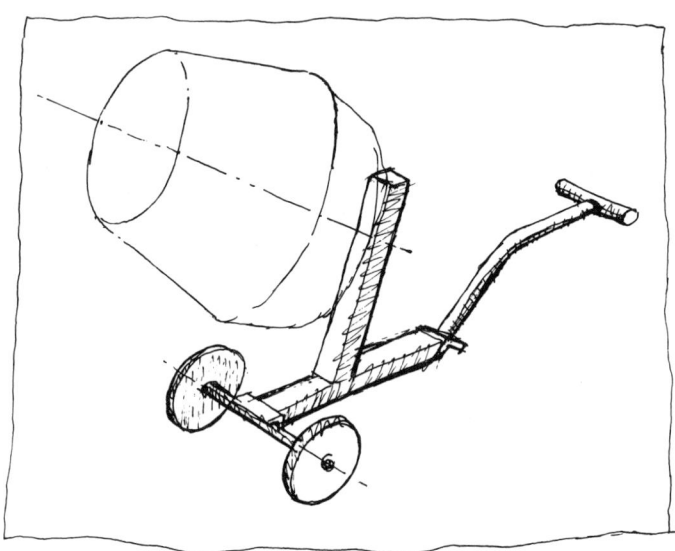

Synthesis, Comparison, and Evaluation of Drive

As specified, a drive is required to reduce from the motor speed of 1450 rev/min to a drum rotation of about 30 rev/min.

$$\text{Required speed reduction} = \frac{1450}{30} = 48.3{:}1$$

Fig. 10.7 shows some possible solutions.

Drive (a) is a common arrangement for this type of equipment, but the high-quality wormbox was considered an unnecessary expense for the hours duration here.

Drive (b) is heavy, bulky, and expensive.

Drive (c) is the favoured arrangement due to its simplicity, neatness, and low cost. Another attractive feature is the spur pinion-shaft and ring gear, for which a standard assembly used on other Benford equipment can (hopefully) be utilised.

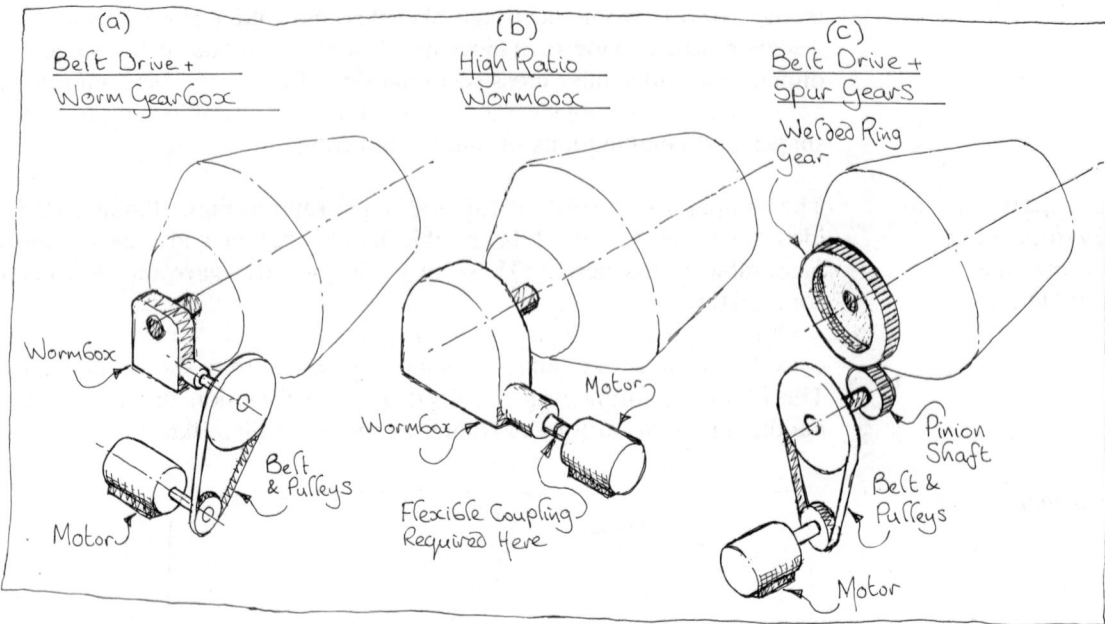

Fig 10.7

Using the standard 11.5:1 gearset and a 4:1 belt-pulley ratio:

Overall ratio $= 11.5 \times 4 = 46:1$

Drum speed $= \dfrac{1450}{46} = 31.5$ rev/min (acceptable)

In similar systematic fashion, the simple, low-cost schemes shown in Fig. 10.8 were chosen for the drum support assembly, pinion-shaft support, and motor mounting, whose basic plate construction also made the frame more rigid. Belt tensioning is provided via slots in the motor frame.

**10.4
Design
Calculations**

**Length of
Tilt-Handle**

Fig. 10.9 shows estimated frame dimensions, total estimated full weight, and the comfortable lifting force of a 5th percentile man (obtained from anthropometric data charts) at 315 mm handle height.

Taking moments about wheel axle:

$$2 \times 68 \times x = 1400 \times 100$$

$$x = 1029 \text{ mm}$$

$$L = 1029 - 400 - 100 = 529 \text{ mm}$$

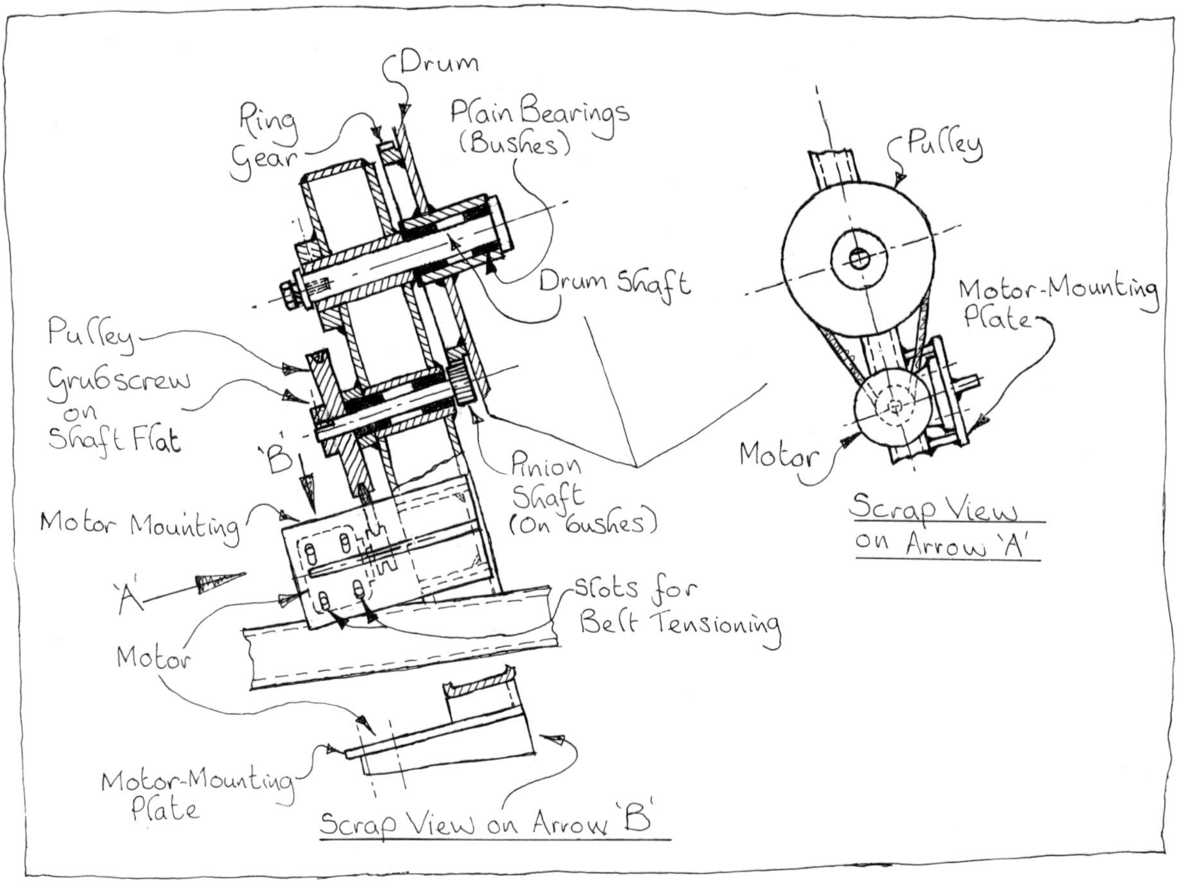

Drum

Ring Gear

Plain Bearings (Bushes)

Drum Shaft

Pulley Grub screw on Shaft Flat

Pulley

Motor-Mounting Plate

'B'

Pinion Shaft (On bushes)

Motor

Motor Mounting

Scrap View on Arrow 'A'

'A'

Slots for Belt Tensioning

Motor

Motor-Mounting Plate

Scrap View on Arrow 'B'

Fig 10.8

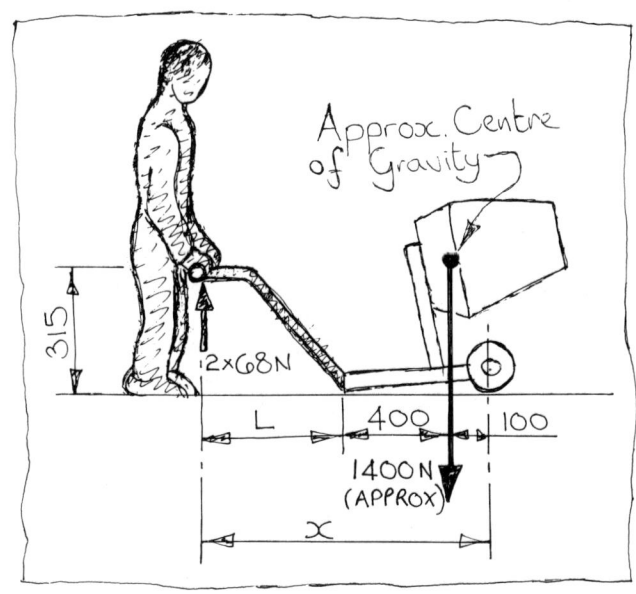

Approx. Centre of Gravity

315

2×68N

L

400

100

1400N (APPROX)

x

Fig 10.9

Size of Handle Tube

This is liable to bend at the base when tilting. Fig. 10.10 shows the calculated length of 614 mm at which the resolved part of the lifting force causes the maximum bending moment.

Max. bending moment $M = 124 \cos 35° \times 614 = 62\,367$ N mm
Assuming BS 4360 Structural Steel Grade 43/A
Tensile/compressive strength of steel $= 460$ N/mm^2

(see p. 101 and chart 2.)

Required safety factor $= 10$

Allowable bending stress $\sigma = \dfrac{460}{10} = 46$ N/mm^2

$$\frac{M}{I} = \frac{\sigma}{y}$$

$$\therefore \frac{I}{y} \text{ (elastic modulus)} = \frac{M}{\sigma}$$

$$= \frac{62\,367}{46} = 1356 \text{ mm}^3$$

Required elastic modulus value $= 1.356$ cm^3

Referring to BS 4848 chart 5, a tube of outside diameter 33.7 mm by wall thickness 2.6 mm has an elastic modulus of 1.84 cm^3 and thus gives a suitable solution. This size of tube is also acceptable for a human hand (see anthropometric data in chapter 6).

Size of Drum Shaft

As the drum shaft does not transmit torque, its size was calculated purely on bending as in the case of the tilt-handle. Fig. 10.11 shows the resolved part of the estimated full drum load creating a dynamic cantilever.

Bending stress calculations thus revealed a minimum allowable shaft diameter of 25 mm using 220M07 (EN8) steel.

Pinion Shaft

As already stated, a standard pinion-shaft was available. This was investigated for combined bending and torsion under fatigue loading.

Pinion details   12 teeth, 2.5 mm module, 20° pressure angle
tooth fillet radius $r_0 = 1$ mm
Required safety factor (on fatigue limit) $= 5$
Material 220M07 (EN8) steel
Tensile strength $= 620$ N/mm^2 (see chart 2)
Bending fatigue limit $= 0.4 \times 620$ (see p. 85)
$= 248$ N/mm^2
Shear strength $= 370$ N/mm^2 (see chart 2)
Torsional fatigue limit $= 0.4 \times 370$ (see p. 85)
$= 148$ N/mm^2

Fig. 10.12 shows the shaft dimensions and locates the weakest point, which is at the pinion shoulder.

Bending stress concentration factor $= 1.2$ (see p. 86)

Fig 10.10

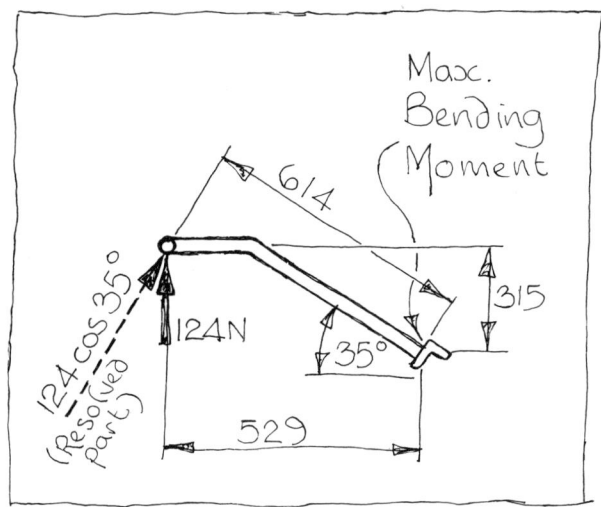

Fig 10.11

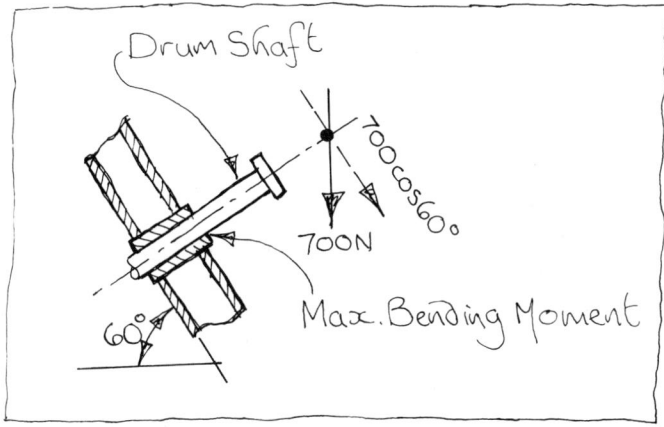

Fig 10.12

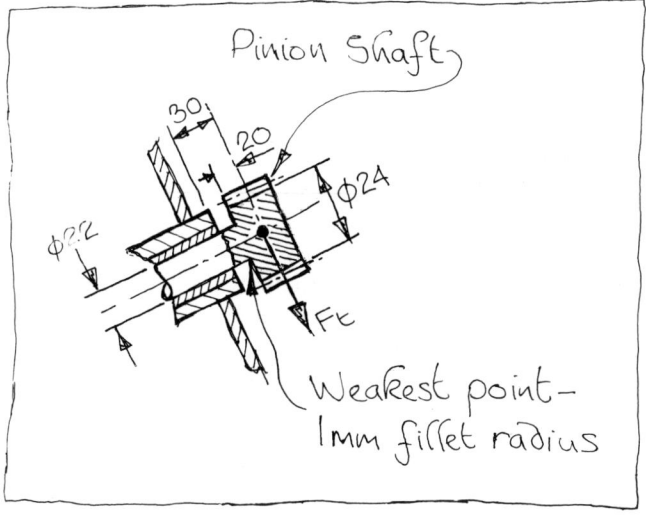

$$\text{Allowable bending stress } \sigma = \frac{248}{1.2 \times 5}$$

$$= 41.33 \text{ N/mm}^2$$

Torsional stress concentration factor $= 1.5$ (see p. 86)

$$\text{Allowable torsion stress } \tau = \frac{148}{1.5 \times 5}$$

$$= 19.73 \text{ N/mm}^2$$

$$\text{Allowable combined stress } q = \sqrt{\left[\left(\frac{\sigma}{2}\right)^2 + \tau^2\right]}$$

$$= \sqrt{\left[\left(\frac{41.33}{2}\right)^2 + (19.73)^2\right]}$$

$$= 28.57 \text{ N/mm}^2$$

$$\text{Speed of pinion } \omega = \frac{1450}{4} = 362.5 \text{ rev/min} = 37.96 \text{ rad/s}$$

$$\text{Torque } T = \frac{P}{\omega} = \frac{250 \text{ watts}}{37.96} = 6.586 \text{ N m} = 6586 \text{ N mm}$$

PCD of pinion $= 12 \times 2.5 = 30$ mm

Pitch circle radius $r = 15$ mm

$$\text{Tangential gear tooth force } F_t = \frac{T}{r} = \frac{6586}{15} = 439 \text{ N}$$

Tooth separating force $F_s = 439 \times \tan 20°$ (see p. 83)

$$= 159.7 \text{ N}$$

$$\text{Resultant tooth force} = \sqrt{[439^2 + 159.7^2]}$$

$$= 467.1 \text{ N}$$

$$\text{Maximum bending moment } M = 467.1 \times 10$$

$$= 4671 \text{ N mm}$$

Equivalent torque $T_e = \sqrt{[M^2 + T^2]}$ (see p. 82)

$$= \sqrt{[4671^2 + 6588^2]}$$

$$= 8076 \text{ N mm}$$

$$\frac{T_e}{J} = \frac{q}{r} \quad \text{(see p. 82)}$$

$$\therefore \frac{8076}{\pi d^4/32} = \frac{28.57}{d/2}$$

$$d = \sqrt[3]{\left[\frac{\frac{1}{2} \times 8076 \times 32}{28.57\pi}\right]}$$

Minimum allowable diameter $= 11.29$ mm

Thus the $\phi 22$ mm standard pinion shaft proved easily adequate for the application.

Gear Capacity

Inspection revealed that the pinion was the weaker of the two mating gears, and that the induction-hardened pinion was more likely to fail in strength than in wear.

Using the Lewis formula quoted on page 98,

$$F_t = Ybc\sigma$$

$$Y = 0.154 - \frac{0.912}{N} \quad \text{(see p. 98)}$$

$$= 0.154 - \frac{0.912}{12}$$

$$= 0.078$$

Circular pitch $C = \pi \times \text{module} = 2.5\pi = 7.854\,\text{mm}$

$$\text{Allowable stress } \sigma = \frac{\text{Tensile or Compressive strength}}{\text{SF} \times \text{SP} \times K_t} \quad \text{(see p. 98)}$$

Safety factor $(\text{SF}) = 5$

$$\text{Speed factor }(\text{SP}) = \frac{3000 + V}{3300} \quad \text{(see p. 98)}$$

$$\text{Pitch line velocity } V = \frac{362.5 \times 30\pi}{60} = 569.4\,\text{mm/s}$$

$$\text{SP} = \frac{3000 + 569.4}{3300} = 1.082$$

Ratio $r_0/C = 1/7.854 = 0.1274$

Stress concentration factor $K_t = 1.6$ (see p.98)

$$\text{Allowable stress } \sigma = \frac{620}{5 \times 1.082 \times 1.6} = 71.63\,\text{N/mm}^2$$

Tangential tooth load $F_t = 439\,\text{N}$

$$\text{Minimum allowable tooth width } b = \frac{F_t}{Y \times C \times \sigma}$$

$$= \frac{439}{0.078 \times 7.854 \times 71.63}$$

$$= 10.01\,\text{mm}$$

The actual tooth width of 20 mm was easily adequate therefore.

Size of Rectangular Tube for Frame

Steel tube to BS 4848 size 60 mm × 40 mm was chosen. This gave an adequate width to support the drum shaft assembly and fitted fairly neatly with the size of the handle. The structure was then investigated for bending and torsion with the critical area located as the base of the upright pedestal as shown in Fig. 10.13.

Fig 10.13

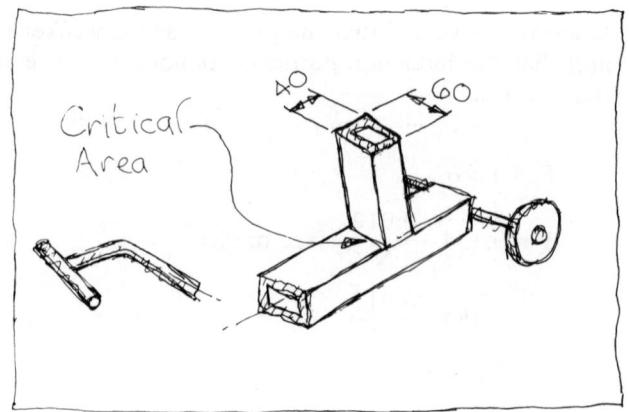

Fig 10.14

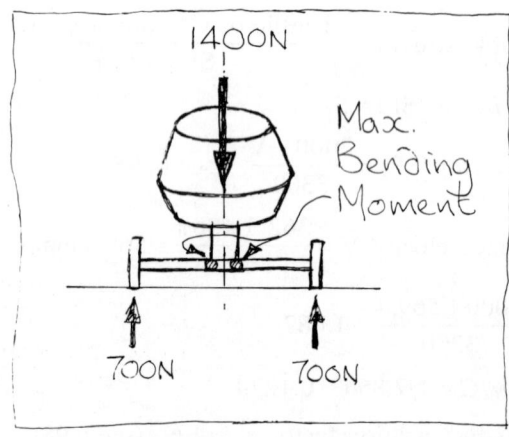

Wheel Axle

Circular steel tube was used and the size calculated for bending in the same manner as the tilt-handle (see Fig. 10.14).

**10.5
Selection of
Standard
Components**

Belts,
Pulleys and
Bearings

These were selected from information in Manufacturers' catalogues after completion of the design assembly drawing; they were then bought-out complete.

Bolts and
Screws

Sizes were determined for tensile loading on the handle bolts and shear loading on the shaft-retaining grub screws.

**10.6
Cost Analysis**

Cost Analysis
of Drum

During the design stages discussed, an engineer from the Estimating Department compared the cost of producing the drum by plastic moulding with that of spun steel.

Using a procedure similar to that described in Chapter 3 it was predicted that a plastic drum would become the cheaper alternative after a break-even quantity of 8000 units produced per year was exceeded. Since it was planned to produce only 5000 units per year, the spun-steel drum was chosen as the cheaper design.

Cost Analysis of Wheels

Using the same analysis procedure as for the drum, it was decided to use plastic wheels as the cheaper alternative to steel pressings.

Overall Cost Analysis

After the completion of the design assembly drawing, a print was sent to the estimating department, who then determined the overall cost per unit.

The investigation revealed a total unit cost of £110, excluding the fixed cost of £200 000 per batch.

For 5000 units, total cost $= £200\,000 + (110 \times 5000)$

$$= £750\,000 \text{ per year}$$

For £160 selling price, total income $= 5000 \times £160$

$$= £800\,000 \text{ per year}$$

Thus there could be expected a slight profit of £50 000 in the first year with a break-even quantity of 4000 units (see Chapter 3). The profit could be expected to increase in following years if fixed costs remained comparatively stable.

**10.7
Re-design for Cost**

Although the overall cost gave a satisfactory result, the following suggestions were put forward by the Estimating Department to reduce cost and thus increase profit margins.

Maximum production reduces component costs and therefore it is essential to quantify the market potential, and the critical factor in this is the selling price.

The main competitors' products are selling at list prices as follows:

	£
Competitor A	184
Competitor B	182
Competitor C	150
Competitor D	260
Competitor E	140

All of the above sell at prices below list price. A and C give up to 30% discount for multiple orders. B and D sell at 15% below list price, whilst D and E average $12\frac{1}{2}\%$ below list price.

True comparisons are therefore:

	£
Competitor A	129
Competitor B	155
Competitor C	105
Competitor D	227
Competitor E	122

which means that in order to compete on similar terms with C we would have to be prepared to discount our proposed list price by 34%, at least in the early days of selling, before our product quality would play a part in the buying decision.

As our unit manufacturing cost is £110 it is necessary to consider a reduction in quality of the unit to make it a viable selling commodity or, alternatively, to provide support from other funds for the product to be initially marketed. The former could be achieved by introducing a less robust drum—reducing steel thicknesses in common with A, C and E, as well as reducing the quality of the electric motor.

There is the final consideration that the Benford name for quality would encourage sales, whereas reducing quality as above may compromise sales of other products.

10.8 Re-design for Function

After the manufacture of the designed items and the arrival of bought-out items, the prototype mixer was assembled in the Research and Development Department and tested as specified.

As a result, the following minor modifications were suggested to improve performance of the product:

1 Extra flat washers on motor mountings to prevent their loosening during operation.
2 Adjustment of 2° to the blade angle to improve discharge properties.
3 Increased diameter of the bottom pulley grub screws to reduce likelihood of slip.
4 Altered rear legs of stand to increase angle in order to improve stability, since there is a tendency to tip backwards if dropped into the rest position after discharge.
5 Torsional deflection of the motor-mounting plate caused excessive vibration under dynamic loading and eventually resulted in failure of electric motor bearings. The mounting-plate was thus re-designed with two simple flanges, which proved to be satisfactory under test. The slight increase in materials cost was balanced by a reduction in manufacturing cost. The re-designed mounting is shown in Fig. 10.15.

Mixer Stand (optional extra)

Fig. 10.16 shows a stand being used for tipping the concrete directly into a wheelbarrow. This was designed from tubular steel and supplied as an optional extra. Figs. 10.17a,b,c,d are another extract from the instruction booklet which illustrates the importance of ergonomic and anthropometric considerations throughout the design project.

10.9 The Finished Product

Figs. 10.18 a,b,c show photographs of the finished mixer design fully assembled, with uncovered drive, and dismantled-for-transport.

For 10.19 shows the specified small cardboard carton in which the complete mixer is packaged in dismantled form. Also in the carton, as specified, is the simple tool kit for easy assembly, and the instruction booklet.

Fig. 10.20 is an exploded projection and parts list for the complete mixer as shown in the instruction booklet. Fig. 10.21 shows every item in the dismantled kit.

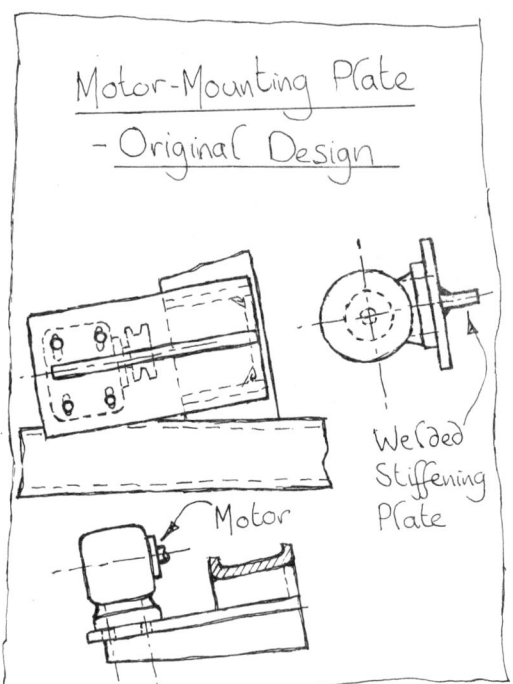
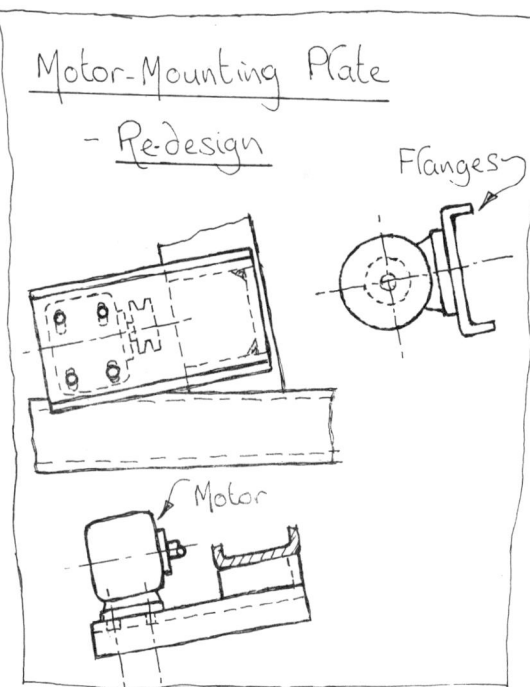

Fig 10.15

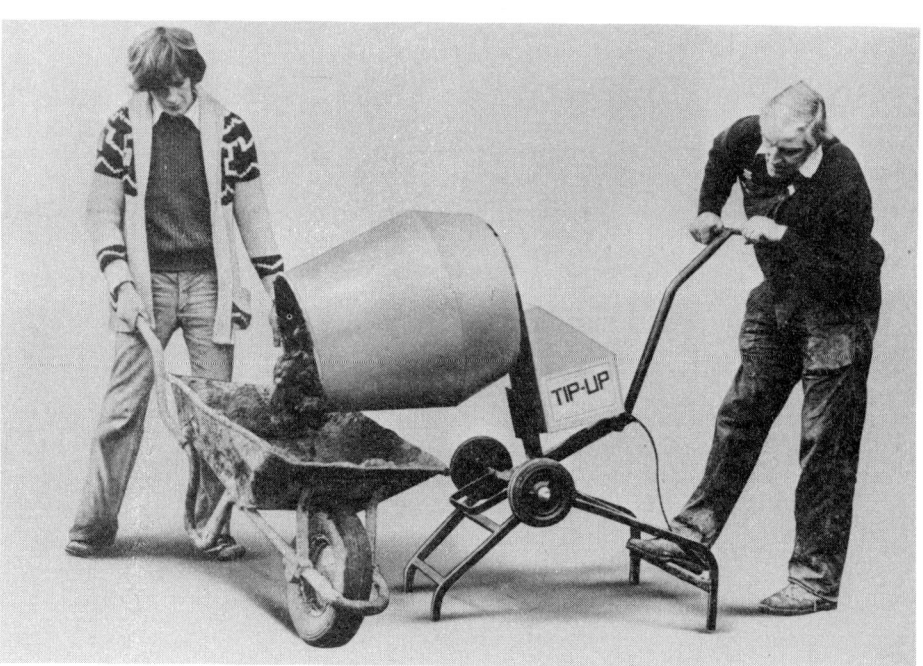

Fig 10.16

Fig 10.17*a*

Fig 10.17*b*

Fig 10.17c

Fig 10.17d

Fig 10.18*a*

Fig 10.18*b*

Fig 10.18*c*

Fig 10.19

Fig 10.20

PARTS LIST

Description	Qty.
Mixer frame	1
Bush	2
Rear drum shield	1
Self tapping screw	4
Pinion shaft	1
Driven pulley	1
Grub screw	2
Motor pulley	1
Electric motor	1
Motor shaft	1
Hex screw	8
Spring washer	8
Hex nut	8
'Vee' belt	1
Motor cover	1
Tipping handle	1
Plastic handgrip	2
Wheel	2
Starlock washer	2
Mixing drum	1
Drum shaft	1
Bush	2
Thrust washer	1
Drum cap	1
Grommet	1
Allen key	1
Spanner	2
Washer	1
Fitting cord	1
Tubular stand	1
Rubber stop	4

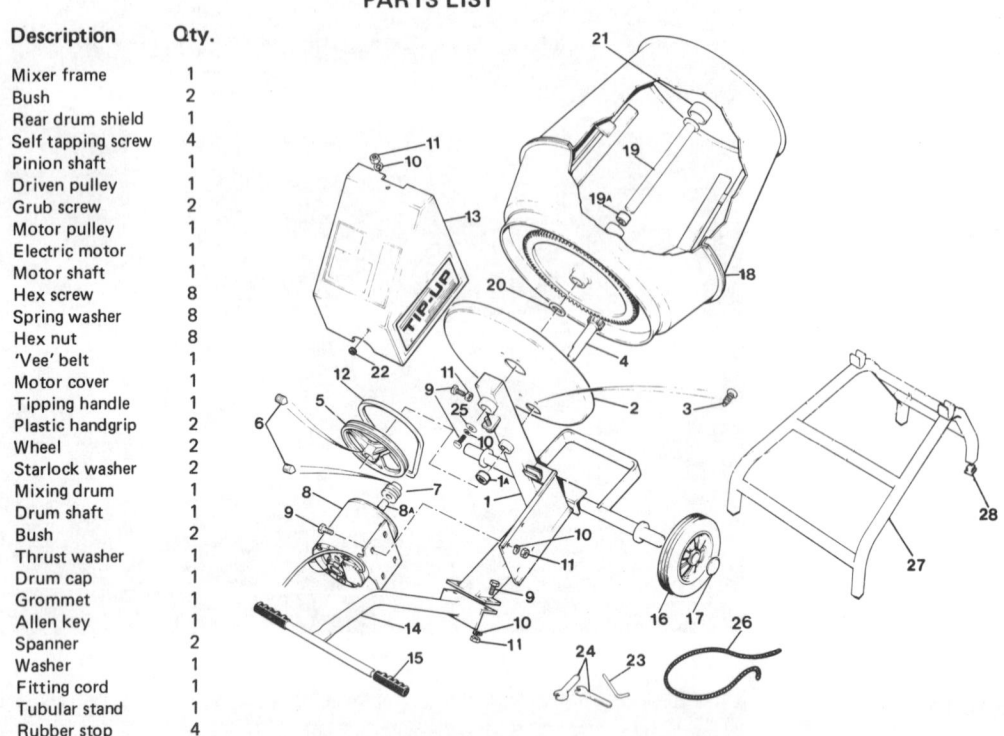

Fig 10.21

11 Assignments

The design assignments suggested here are of varying complexity, with due regard to time allowance and number of students involved in each individual assignment.

If ergonomic and anthropometric information is required, much of this can be obtained from Chapter 6. For further information, the authors recommend reference to literature by Messrs. E. Grandjean, John Croney, K. F. H. Murrell, and to B.S.I. document PD 7302.

Assignment Reports

Where a team assignment is undertaken, it is recommended that each student submits a full individual report, neatly presented (preferably typed) and supplied in an indexed folder.

The report content will vary with the particular requirements of each assignment, but general guidelines are shown in Chapter 10, which describes the complete design case study of a small concrete mixer.

Main headings of the report may include:

a) *Introduction.* (This may include the assignment requirements, and preliminary events leading to their formulation in a typical company.)

b) *Complete design specification.*

c) *Planning schedules.*

d) *Synthesis, comparison and evaluation.* (Any sketches of design ideas, photographs and notes on scale models produced, and description of systematic analysis to be included.)

e) *Cost analysis.*

f) *Calculations.*

g) *Anthropometric and aesthetic considerations.* (Notes, sketches, manikin models, etc.)

h) *Design general arrangement drawing.* (To include a parts list, any required scrap views, and notes on required materials and production techniques. It is recommended that the drawing is supplied as a dyeline print of at least A2 size and neatly folded to fit the A4 folder.)

i) *Development of design.* (Description of likely aspects of testing and possible areas of re-design.)

j) *Conclusion.* (Limitations and problems experienced in the assignment. Likely future developments for a typical company.)

**Assignment 1
Cordless Hand
Vacuum Cleaner**

A company wishes to produce a cordless vacuum cleaner, suitable for use on car upholstery or small domestic spillages. The device is to be provided by a re-chargeable battery to give a continuous use of up to 15 minutes.

The drive is to be via a small 1500 rev/min electric motor purchased as a standard component.

Market research indicates the required production quantity to be in the region of 10 000 per year.

The product should be light, compact and of suitable size and shape to suit the anthropometric requirements of 90% of the adult male and female population (see Chapter 6).

It should also be safe and have a pleasing aesthetic appeal (see Chapter 8).

a) Complete an appropriate specification for the vacuum cleaner using examples shown in Chapters 1 and 10 for your guidelines.
b) Sketch design schemes of
   i) At least three proposals for the outer casing/handle.
   ii) At least two proposals for the drive assembly from motor to output rotor blades which have a speed of approximately 200 rev/min.
In each case choose one of the proposals in a systematic manner as described in Chapters 1 and 10.
c) Conduct a break-even cost analysis exercise (see Chapters 3 and 10), comparing the following methods of producing the outer-casing/handle:
   i) Plastic injection moulding.
   ii) Zinc-alloy die-casting.
Use realistic values for set-up cost and unit cost of manufacture for each production technique.
d) Complete a network analysis planning schedule for the design and development and production activities required for the completed product.
e) List any areas of re-design which may be required for function or cost.
f) Draw a complete design assembly drawing of the vacuum cleaner, including a parts list notes regarding recommended materials and production methods.

The following British Standards may be of assistance in this assignment.

BS 3456 Specification for safety of household and similar electrical appliances, section 3.3.
BS 3999 Method of measuring performance of household electrical appliances.
BS 5415 Specification for safety of electrical motor operated industrial cleaning devices.

**Assignment 2
Centrifugal
Clutch**

Fig. 11.1 shows the output shaft A of a flange-mounted centrifugal clutch which is required to engage with the output shaft B of a lawn-mower petrol engine. This engine transmits a power of 2 kW at 1800 rev/min through the clutch and finally to the cutting blades via a 4:1 vee-belt reduction system, whose overhung driver pulley C is attached to shaft A.

Fig 11.1

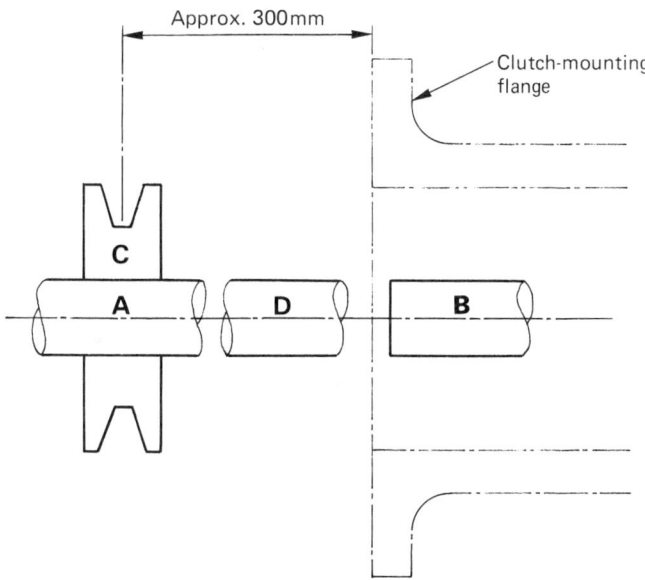

The centrifugal clutch is required to engage at a speed of 900 rev/min with a maximum pressure of 18 kPa between rubbing surfaces of approximately 0.35 coefficient of friction.

Shaft B is directly coupled to the clutch input shaft D.

Both shafts A and D are to be supported on roller and/or ball bearings.

a) Neatly sketch at least two design proposals for the centrifugal clutch mechanism and choose one of these using a systematic procedure as outlined in Chapter 1.

b) Decide on a suitable area of rubbing surfaces for the chosen scheme and thus determine the required normal force between frictional surfaces. Hence determine the required radius of rubbing surfaces for the transmitted torque. From this radius and chosen width of rubbing surfaces, determine approximate overall dimensions for the assembly and approximate positions of shaft bearings.

c) From a manufacturers catalogue select a standard vee-belt pulley for item (c), following clearly the procedure outlined in the catalogue.

d) Choosing suitable materials (see charts 1, 2, 3) determine the minimum diameters of shafts A and B on the basis of torsional fatigue and, when necessary, bending fatigue. (Assume a safety factor of 3 on fatigue limit.) (Refer to Chapters 5 and 10 for guidance.)

e) Select suitable shaft bearings using the procedure outlined in Chapter 5 and complete fully-dimensioned detail sketches of shafts A and B, taking due consideration of fatigue stress concentration factors as outlined in Chapter 5.

f) Complete two sectional general assembly views of the complete clutch arrangement including all necessary keys, bearing locations, oil seals and threaded fastenings. Cross-reference all items shown to a parts list which will include selected materials.

**Assignment 3
Cycle Exerciser**

It is required to design and produce a dual-action cycle exerciser for use in the home or a gymnasium. The apparatus should incorporate a variable-tension control on the pedal mechanism and a device to give an oscillating movement of the handlebars in conjunction with the pedal rotation as shown in Fig. 11.2.

Fig 11.2

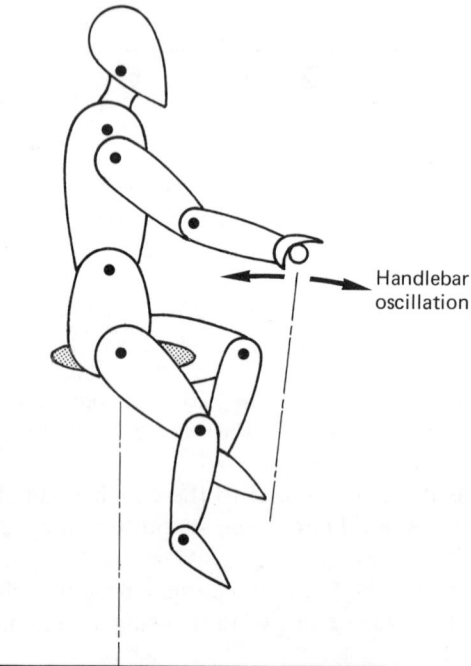

Handlebar oscillation

The design should be light, compact, portable and preferably foldable for easy storage, but must also be safe, stable and sufficiently rigid to comfortably support 90% of the male and female adult population. With due regard to its domestic market, the product should also incorporate good aesthetic appeal as outlined in Chapter 8.

Estimated quantities are in the region of 4000 per year.

a) With reference to the anthropometric data in Table 6.1, construct a number of two-dimensional pin-jointed cardboard manikin scale models of the human form to determine the critical dimensions of the exerciser which will most favourably accommodate the required range of population. If necessary, provision should be made for height adjustment of saddle and handlebars.

b) List any further ergonomic and anthropometric requirements of the design.

c) Compile a complete design specification for the product along the lines suggested in Chapters 1 and 10.

d) Conduct a systematic analysis exercise with at least three design proposals for the supporting framework, and at least two design proposals for each of the pedal tension control and the handlebar oscillation (see Chapters 1 and 10).

e) Using suitable materials and safety factors, determine the minimum standard size of cross-section for all critically-loaded features.

f) Draw a complete design assembly drawing of the exerciser, including a parts list and notes regarding recommended materials and production methods.

**Assignment 4
Pneumatic
Robot Arm**

A pneumatically-powered robot arm is required for a wide variety of applications similar to that shown in Fig. 11.3.

The mass-produced shouldered bush is required to be picked up from position A and placed as shown at position B.

The sequence of operations is as follows:

1) Jaws clamp the bush.
2) Arm rises 100 mm vertically.
3) Arm swings through 90° rotation.
4) Arm extended linearly by 160 mm to position B.
5) Jaws rotate 180° about the arm axis to place the bush as shown at position B.

Neatly sketch at least two design proposals for converting linear pneumatic action into each of actions 1), 2) and 5). In each case choose a "best solution" using a systematic procedure similar to that shown in Chapter 1.

Draw a complete general assembly of your design including any necessary scrap views of sub-assemblies. Complete a parts list for all items shown, including indication of materials used.

The completed design assembly should be robust, compact, and easily transported as a single unit, and should have a simple layout with a minimum possible number of separate pneumatic linear activities.

Provision should be made to adjust the radius of arm swing for different applications.

The number of components produced are expected to be in the region of 5000 per year.

The following British Standards may be of assistance in this assignment:

BS 5791 Pneumatic tools and machines glossary.
BS 3512 Testing of pneumatic transmitters.
BS 4862 Pneumatic cylinders.
BS 5409 Pneumatic installation—nylon tubing (Part 1).

Fig 11.3

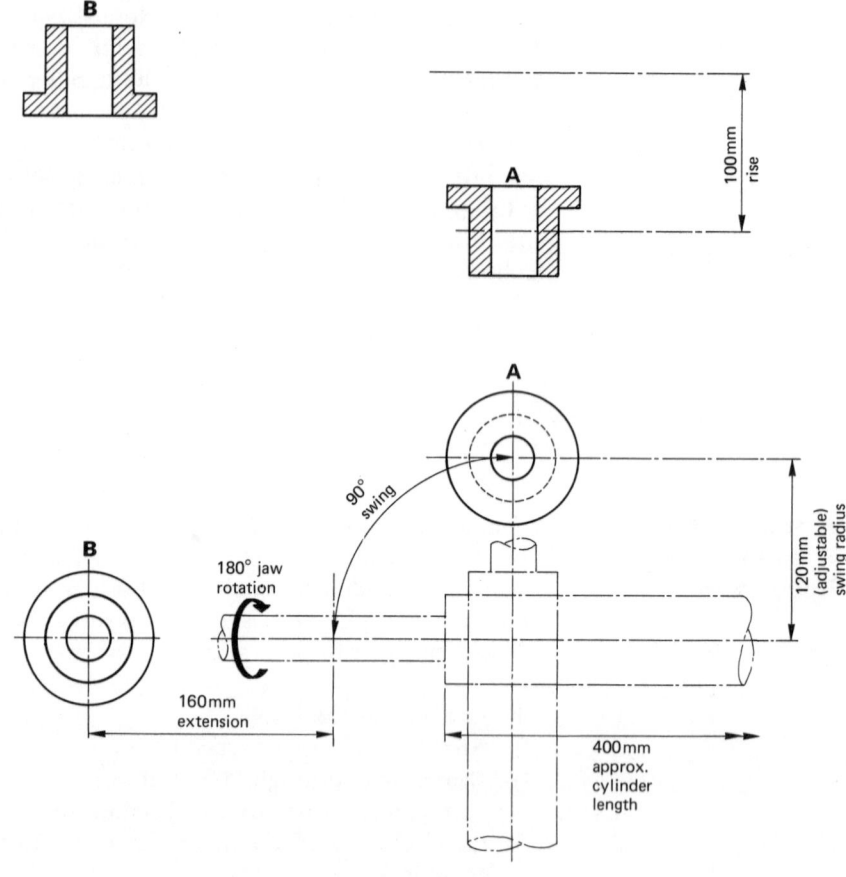

**Assignment 5
Hand Fishing-net
Machine**

This assignment is included by kind permission of the Education Department, OXFAM, and is an extract from one of their design problem packs concerning "Intermediate Technology" for third-world countries.

Specific design tasks will not be outlined, but the problem would appropriately be tackled as a group assignment and presented along the lines described at the beginning of this chapter.

Fishing nets in many developing countries are still made by hand, thus sustaining employment for the fishermen at times when they are not fishing. This is, however, an extremely laborious process, and fishermen are being drawn into buying imported nets because of the time involved in making their own. This situation has highlighted the need for a simple hand-operated fishing-net machine that can be used by individual fishermen or small co-operatives to manufacture their own nets.

The options in net manufacturing are not very good for the fishermen: it is either completely by hand, or by fully automatic factory machinery (the cheapest available in the U.K. costing well over £30 000), and because of the method of operation these machines are as long as the net is wide.

What seems to be needed is a machine something like a sewing machine which will allow the net to pass on through it after each knot is formed.

There would be no problem (for example) in leaving the finishing of the ends to hand labour.

Ideally the machine must be able to accommodate a variety of twine and mesh sizes, and must produce symmetrical meshes.

Some starting ideas are given below.

Starting Ideas

1 Some type of machine that could feed along after the first line and row of "loops" have been put up.

2 Look at the old bailing machines used by farmers as these "knot" the string.

3 *a*) Look at traditional net manufacture.

b) Look at commercial net manufacture.

Figs. 11.4*b* and 11.4*c* show some traditional knot patterns. Fig. 11.4*d* illustrates the common types of net needle.

Fig 11.4

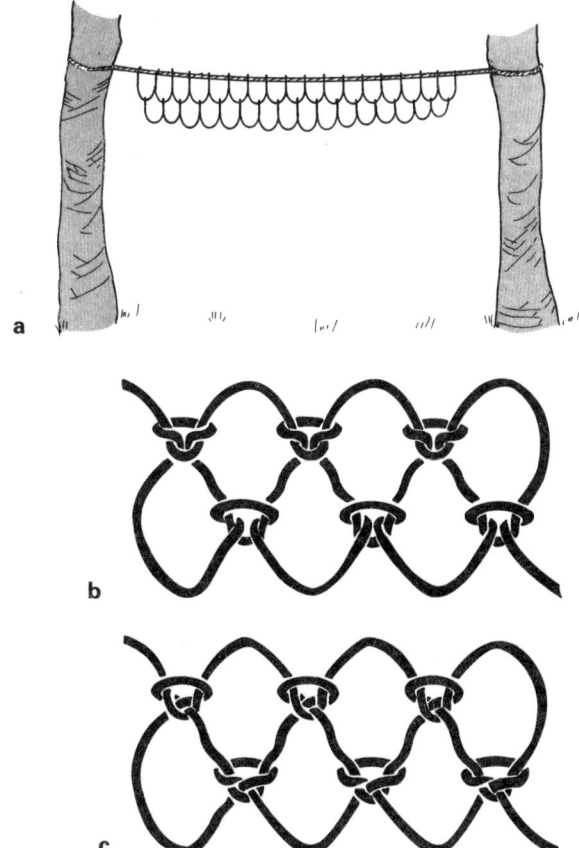

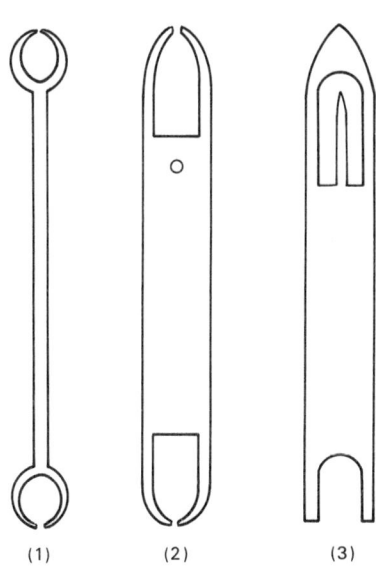

The usual types of net needles:

(1) Filet needle of the Mediterranean area

(2) Filet needle of Southern Germany

(3) Modern needle with tongue

d

**Assignment 6
Multi-purpose
Child's Seat**

A company has decided to design and produce a convertible seating arrangement for a child of approximately six months to three years old.

A standard seat with safety harness is required to attach to the following alternative fittings:

i) A pushchair which may be easily detached from the standard seat and folded to a suitable compact shape for placing in the boot of a family car.

ii) Strap fittings with suitable pattern, strength and fastenings to provide a suitable child's car-seat attached to the back seat and floor of the car.

iii) A "high chair" arrangement for the child during family meals at the table. This is to include an attachment for supporting food and utensils.

iv) An "automatic swing" arrangement, preferably using the same framework as the high chair. This is to incorporate a clockwork mechanism to safely swing the child at about 25 oscillations per minute for a period of approximately ten minutes.

Complete such a design assignment to include:

a) A complete design specification for the multi-purpose seating arrangement.

b) Systematic analysis schemes as outlined in Chapters 1 and 10, comprising at least two alternative proposals for the framework of the folding pushchair and at least three alternative proposals for the clockwork mechanism on the automatic swing.

c) Any relevant calculations required for structural section sizes, using appropriate chosen materials.

d) Separate general assembly drawings for each of the alternative arrangements, including all necessary scrap views of sub-assemblies. All items are to be called up on a parts list. Choice of materials and brief description of production techniques are to be included in note form on the drawings.

e) A choice of suitable selling price based on market investigations and realistic material and production costs, for an estimated production quantity of 8000 per year.

The design is to be ergonomically and anthropometrically suited to 90% of the child population.

All features of the design are to conform to relevant requirements of the following British Standards and legislation:

BS 4792 Safety requirements for pushchairs.

Statutory Instrument 1978, No. 1372 Consumer protection. perambulator and pushchair safety regulations.

BS 3254 Seatbelt assemblies for motor vehicles.

AU 157 (BSI document) Childrens restraining devices in automobiles.

BS 5799: Parts 1 & 2, Childrens high chairs and multi-purpose chairs for domestic use.

BS 5696 (concerning play equipment).

BS 5665: Part 1, Safety of toys.

**Assignment 7
Land Rover
Breakdown Crane**

It is required to design a hand-operated mechanical crane to fit on the back of a Land Rover used in vehicle breakdown service. The crane is to conform approximately to the dimensions shown in Fig. 11.5 and should be able to lift at two wheels and tow vehicles up to the size of a large family estate car.

Study Chapter 10 which describes the complete design case study of a small cement mixer, and conduct a similar exercise for the required crane.

The design project should include:

a) A realistic design specification on the lines of the examples in Chapters 1 and 10, including reference to market research conducted and quantity forecasts.

b) A list of the design activities required to produce a working prototype and a critical path network analysis as outlined in Chapters 2 and 10.

c) The construction of balsa wood scale models of at least three design schemes for the crane framework structure and a choice of best solution along similar lines to the Bridge Project in Chapter 7, the choice being made mainly on results from qualitative tests for stiffness/mass ratios under relevant identical loading conditions.

d) A break-even cost analysis exercise comparing two feasible production techniques for at least one main component of the design (see Chapter 3).

e) A systematic design exercise along the lines described in Chapter 1 to choose a best solution from at least three design proposals for each of the following:

    i) A mechanism supplying adequate mechanical advantage to lift the range of vehicles required via a crank handle operated by a 5th percentile man. Force calculations should be shown.

    ii) A reversible ratchet/clutch mechanism which allows safe lifting or lowering as required.

f) A device to prevent vehicles from swinging sideways while being towed.

g) Economic choice of steel section for the crane framework, giving valid reasons (see Chapter 5) and all necessary design calculations to obtain the minimum allowable standard section for the loadings executed using a suitable safety factor (see Chapter 5 and charts 4 and 5).

Fig 11.5

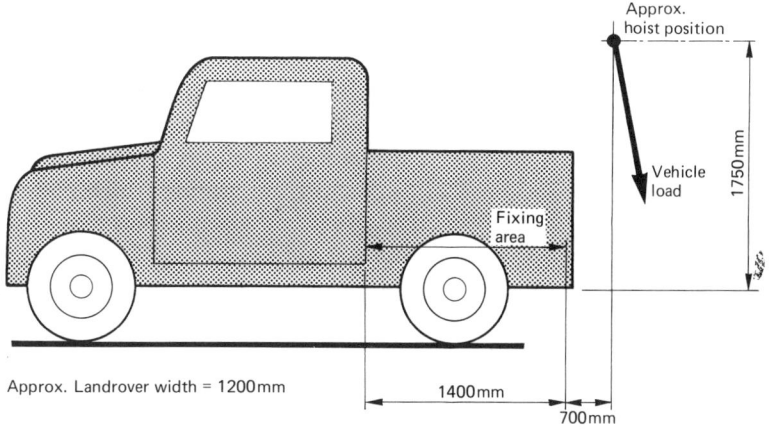

h) A general assembly drawing of the complete layout including all necessary scrap views of sub-assemblies. All items are to be called up on a parts list.

Choice of materials and brief description of production techniques are to be included in note form on the drawing.

The following British Standards may be of assistance in this assignment:
BS 2573: Part 1 (1977) Specification for the permissible stresses in cranes and design rules.
BS 3810: Part 4 (1968) Terms used in connection with cranes.
CP 3010 (1972) Safe use of cranes.

**Assignment 8
Exercise in
Brainstorming—
Group Assignment**

A company has decided to produce a new design of trailer for use behind a family car. This is to be capable of transporting camping equipment or any small domestic loads.

The Chief Engineer has decided to form a committee to design the basic concept of the new trailer. The committee is to consist of members of staff from the Design Office, the Production Engineering Department and the Sales and Marketing Departments.

Basic Specification
The committee should consider the following items in the order given.
1) Lightweight frame.
2) Low profile.
3) Small wheels.
4) Rear tailboard.
5) Cover over top.
6) Shape of mudguards.
7) Material of side panels.

Method and Rules
1. The committee is to use the basic fundamentals of brainstorming whereby all ideas are to be written down as any committee member calls them out. Do not try to sort out the ideas at this stage.
2. Consider each of the items of the basic specification one by one.
3. Move on to the next item when it is considered that ideas have been exhausted.
4. To enable each member of the committee to recognise the others role, place a piece of paper in front stating which department they are from.

Brainstorming Team
This is to consist of
2 Designers
2 Production Engineers
1 Sales Engineer
1 Marketing Person.

Observers
The remaining students act as observers and note the following:

    a) Were all the objectives achieved. If not, why?

    b) How did noise level vary from the start, and how did it affect the decisions made?

    c) Which department put forward most ideas?

    d) What were the reactions of each team member to operating without a chairman? Was there a tendency for one person to try and chair the meeting?

    e) Did any idea prompt other team members to put forward better solutions in great quantity?

    f) What were your general impressions of the effectiveness of this method.

The Committee Member's Roles
The Designer: concerned only with the design features of the trailer and should not form an interest in any other aspect.
The Production Engineer: concerned only with those design features which are associated with manufacturing methods.
The Sales Engineer: confined to putting forward ideas affecting selling, for example appearance and size.
The Marketing Person: concentrates only on putting forward ideas on design which concern cost, appearance, shape and size.

**Assignment 9
Exercise in
Ergonomics:
Bearing Bush
Assembly**

It is required to assemble a bearing bush into a casting bore (Fig. 11.6) and to carry out subsequent operations thus:

    i) Machine bush bore in situ.

    ii) Inspect and pass completed assembly to the stores.

Fig. 11.6*a* shows a flow diagram illustrating the route which the parts will take until they reach the stores as a complete assembly.

a) Design a suitable clamp for holding the work while drilling the bearing bore in-situ. Fig. 11.6 shows the basic casting with bush pressed in.

b) Sketch the pillar drill shown in Fig. 11.6*b* and insert the dimensions indicated, assuming that the operator is standing.

c) Sketch the hand press shown in Fig. 11.6*c* and insert the dimensions indicated, assuming that the operator is seated.

d) An inspector sits at a metalwork bench to check the final assemblies. Sketch a suitable plug gauge for checking the bore. Sketch the work bench and insert the dimensions indicated in Fig. 11.6*d*. The inspector is in a seated position.

e) Adjacent to the inspector is a storage bin for placing the finished work. Design a suitable bin capable of stacking the finished components, indicating any special stacking arrangement considered necessary.

f) The bin is moved with the aid of a hydraulic trolley. How is this trolley arranged to pick up the bin? Show correct height of controls to operate the lifting device and also the handle for pushing the trolley.

Use the table of anthropometric data (Table 6.1) for establishing the dimensions. Draw the outline of the operator in each sketch and show also the position of the components for ease of handling.

Fig 11.6

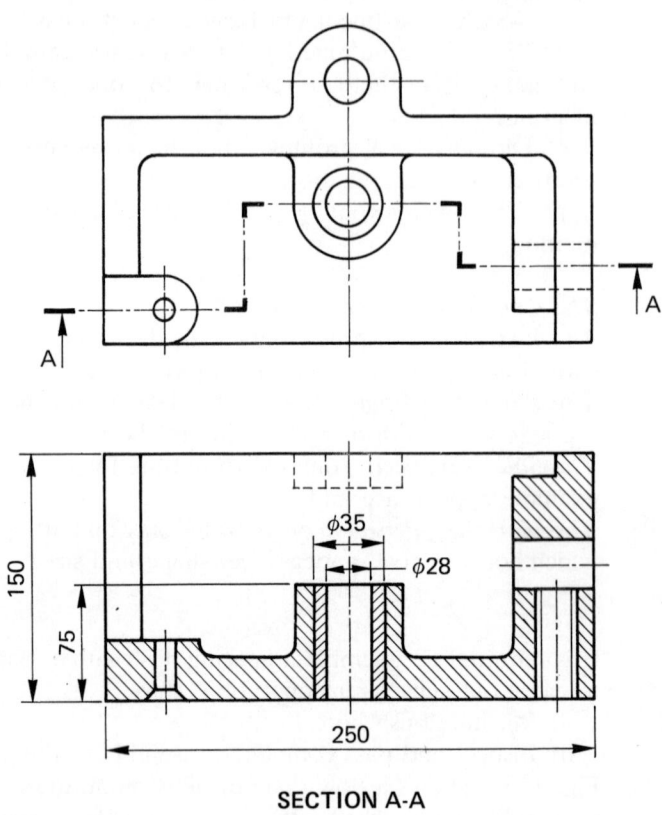

SECTION A-A

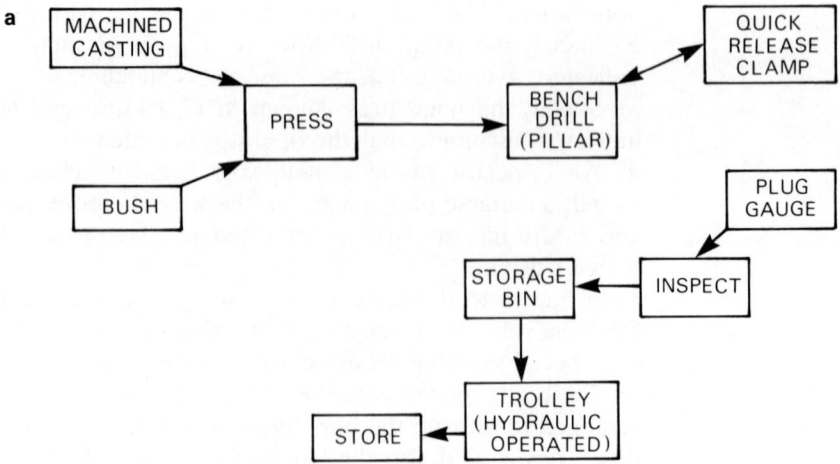

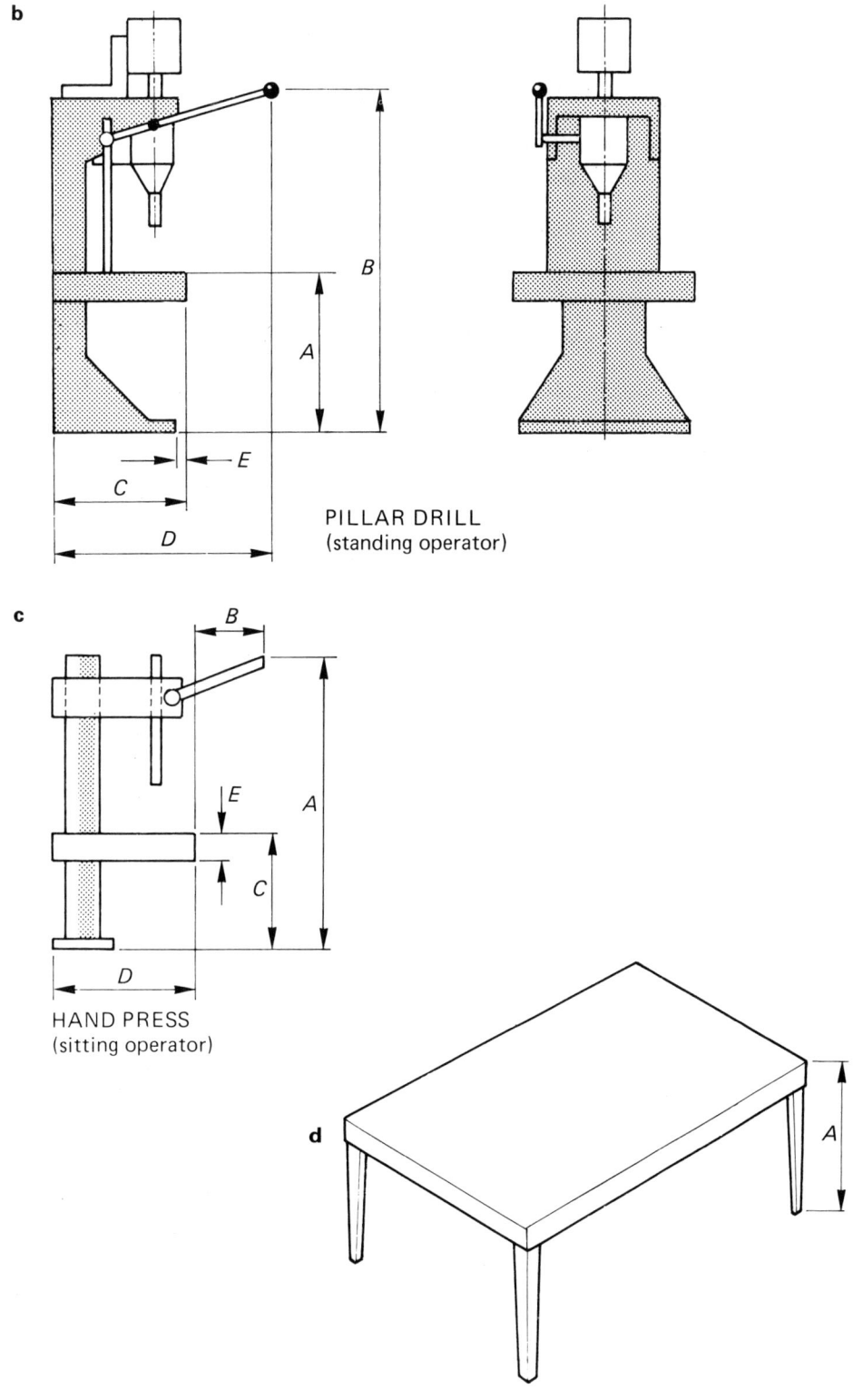

b

PILLAR DRILL
(standing operator)

c

HAND PRESS
(sitting operator)

d

Chart 1

	Grey cast iron (grade 14/17)	Copper	Brass (60/40)	Aluminium	Aluminium alloys		
					LM14	LM6	Dural-umin
Tensile strength (N/mm^2)	200	150	220	100	280	180	400
Compressive strength (N/mm^2)	650	220	80	100	200	100	350
Shear strength (N/mm^2)	110	180	110	60	200	120	250
Density (kg/m^3)	7200	8900	8300	2700	2800	2600	2800
Brinell No.	150	50	65	40	110	70	150
Thermal conductivity (% silver)	13	92	24	35	30	30	30
Electrical conductivity (% silver)	15	94	25	57	30	30	30
Melting point (°C)	1200	1080	900	660	540	560	540
Coefficient of expansion (/°C)$\times 10^{-6}$	11	17	20	24	24	24	24

Chart 2

	Carbon steels					Phosphor bronze	S.G. iron
	070 A72 EN 42	045 M10 EN 32A	080 M40 EN 8	070 M20 EN 3	220 M07 EN 1A		
Tensile strength (N/mm^2)	600	500	620	460	420	350	650
Compressive strength (N/mm^2)	600	500	620	460	420	—	650
Shear strength (N/mm^2)	360	300	370	270	250	200	120
Density (kg/m^3)	7800	7800	7800	7800	7800	8600	7700
Brinell No.	200	140	200	120	110	150	200
Thermal conductivity (% silver)	14	14	14	14	14	40	13
Electrical conductivity (% silver)	10	10	10	10	10	25	15
Melting point (°C)	1400	1500	1450	1500	1500	900	1100
Coefficient of expansion (/°C)$\times 10^{-6}$	12	12	12	12	12	17	11

Chart 3

	Nylon (grade 66)	Zinc alloy	Tin	Stainless steel (18/8)	Nickel-chrome steel 817 M40 (EN 24)	Chromium steel 526 M60 (EN 11)	High speed steel
Tensile strength (N/mm^2)	80	200	15	600	1200	900	—
Compressive strength (N/mm^2)	180	150	—	—	1200	900	—
Shear strength (N/mm^2)	—	100	—	360	720	540	—
Density (kg/m^3)	1200	7100	7300	7900	7900	7900	7900
Brinell No.	25	80	5	200	340	700	660
Thermal conductivity (% silver)	—	26	16	14	14	14	14
Electrical conductivity (% silver)	—	27	14	2	10	10	10
Melting point (°C)	240	390	230	1500	1500	1500	1400
Coefficient of expansion (/°C)$\times 10^{-6}$	100	27	20	16	12	12	12

Chart 4a

CHANNELS

To BS4: Part 1

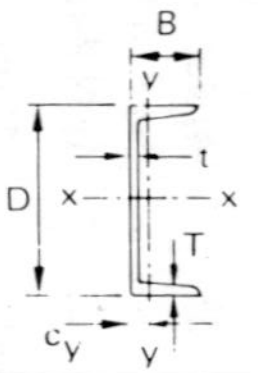

Designation		Depth of section	Width of section	Thickness		Distance	Area of section	Moment of inertia		Radius of gyration		Elastic modulus		Plastic modulus	
Nominal size	Mass per metre	D	B	Web t	Flange T	c_y		Axis x-x	Axis y-y	Axis x-x	Axis y-y	Axis x-x	Axis y-y	Axis x-x	Axis y-y
mm	kg	mm	mm	mm	mm	cm	cm^2	cm^4	cm^4	cm	cm	cm^3	cm^3	cm^3	cm^3
76 × 38	6.70	76.2	38.1	5.1	6.8	1.19	8.53	74.14	10.66	2.95	1.12	19.46	4.07	23.4	7.76
102 × 51	10.42	101.6	50.8	6.1	7.6	1.51	13.28	207.7	29.10	3.95	1.48	40.89	8.16	48.8	15.71
127 × 64	14.90	127.0	63.5	6.4	9.2	1.94	18.98	482.5	67.23	5.04	1.88	75.99	15.25	89.4	29.31
152 × 76	17.88	152.4	76.2	6.4	9.0	2.21	22.77	851.5	113.8	6.12	2.24	111.8	21.05	130.0	41.26
152 × 89	23.84	152.4	88.9	7.1	11.6	2.86	30.36	1166	215.1	6.20	2.66	153.0	35.70	177.7	68.12
178 × 76	20.84	177.8	76.2	6.6	10.3	2.20	26.54	1337	134.0	7.10	2.25	150.4	24.72	175.4	48.07
178 × 89	26.81	177.8	88.9	7.6	12.3	2.76	34.15	1753	241.0	7.16	2.66	197.2	39.29	229.6	75.44
203 × 76	23.82	203.2	76.2	7.1	11.2	2.13	30.34	1950	151.3	8.02	2.23	192.0	27.59	225.2	53.32
203 × 89	29.78	203.2	88.9	8.1	12.9	2.65	37.94	2491	264.4	8.10	2.64	245.2	42.34	286.6	81.62
229 × 76	26.06	228.6	76.2	7.6	11.2	2.00	33.20	2610	158.7	8.87	2.19	228.3	28.22	270.3	54.24
229 × 89	32.76	228.6	88.9	8.6	13.3	2.53	41.73	3387	285.0	9.01	2.61	296.4	44.82	348.4	86.38
254 × 76	28.29	254.0	76.2	8.1	10.9	1.86	36.03	3367	162.6	9.67	2.12	265.1	28.21	317.4	54.14
254 × 89	35.74	254.0	88.9	9.1	13.6	2.42	45.52	4448	302.4	9.88	2.58	350.2	46.70	414.4	89.56
305 × 89	41.69	304.8	88.9	10.2	13.7	2.18	53.11	7061	325.4	11.5	2.48	463.3	48.49	557.1	92.60
305 × 102	46.18	304.8	101.6	10.2	14.8	2.66	58.83	8214	499.5	11.8	2.91	539.0	66.59	638.3	128.1
381 × 102	55.10	381.0	101.6	10.4	16.3	2.52	70.19	14894	579.7	14.6	2.87	781.8	75.86	932.7	144.4
432 × 102	65.54	431.8	101.6	12.2	16.8	2.32	83.49	21399	628.6	16.0	2.74	991.1	80.14	1207	153.1

Chart 4b

JOISTS

To BS4: Part 1

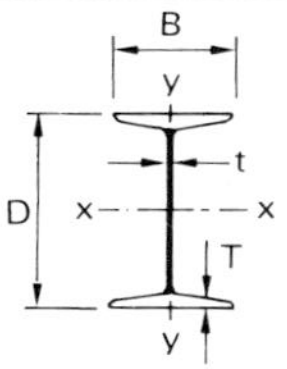

Designation		Depth of section D	Width of section B	Thickness		Area of section	Moment of inertia		Radius of gyration		Elastic modulus		Plastic modulus	
Nominal size	Mass per metre			Web t	Flange T		Axis x-x	Axis y-y	Axis x-x	Axis y-y	Axis x-x	Axis y-y	Axis x-x	Axis y-y
mm	kg	mm	mm	mm	mm	cm^2	cm^4	cm^4	cm	cm	cm^3	cm^3	cm^3	cm^3
254 x 203	81.85	254.0	203.2	10.2	19.9	104.4	12016	2278	10.7	4.67	946.1	224.3	1076	370.4
254 x 114	37.20	254.0	114.3	7.6	12.8	47.4	5092	270.1	10.4	2.39	401.0	47.19	460.0	79.30
203 x 152	52.09	203.2	152.4	8.9	16.5	66.4	4789	813.3	8.48	3.51	471.4	106.7	539.8	175.5
203 x 102	25.33*	203.2	101.6	5.8	10.4	32.3	2294	162.6	8.43	2.25	225.8	32.02	256.3	51.79
178 x 102	21.54*	177.8	101.6	5.3	9.0	27.4	1519	139.2	7.44	2.25	170.9	27.41	193.0	44.48
152 x 127	37.20	152.4	127.0	10.4	13.2	47.5	1818	378.8	6.20	2.82	238.7	59.65	278.6	99.85
152 x 89	17.09*	152.4	88.9	4.9	8.3	21.8	881.1	35.98	6.36	1.99	115.6	19.34	131.0	31.29
152 x 76	17.86	152.4	76.2	5.8	9.6	22.8	873.7	60.77	6.20	1.63	114.7	15.90	132.5	26.67
127 x 114	29.76	127.0	114.3	10.2	11.5	37.3	979.0	241.9	5.12	2.55	154.2	42.32	180.9	70.85
127 x 114	26.79	127.0	114.3	7.4	11.4	34.1	944.8	235.4	5.26	2.63	148.8	41.19	171.9	68.07
127 x 76	16.37	127.0	76.2	5.6	9.6	21.0	569.4	60.35	5.21	1.70	89.66	15.90	103.6	26.28
127 x 76	13.36*	127.0	76.2	4.5	7.6	17.0	475.9	50.18	5.29	1.72	74.94	13.17	85.23	21.29
114 x 114	26.79	114.3	114.3	9.5	10.7	34.4	735.4	223.1	4.62	2.54	128.6	39.00	151.2	65.63
102 x 102	23.07	101.6	101.6	9.5	10.3	29.4	486.1	154.4	4.06	2.29	95.72	30.32	113.4	50.70
102 x 64	9.65*	101.6	63.5	4.1	6.6	12.3	217.6	25.30	4.21	1.43	42.84	7.97	48.98	12.91
102 x 44	7.44	101.6	44.4	4.3	6.1	9.5	152.3	7.91	4.01	0.91	30.02	3.44	35.30	5.99
89 x 89	19.35	88.9	88.9	9.5	9.9	24.9	306.7	101.1	3.51	2.01	69.04	22.78	82.77	38.03
76 x 76	14.67	76.2	80.0	8.9	8.4	19.1	171.9	60.77	3.00	1.78	45.06	15.24	54.16	25.73
76 x 76	12.65	76.2	76.2	5.1	8.4	16.3	158.6	52.03	3.12	1.78	41.62	13.60	48.84	22.51

Note: Joists marked * have a 5° taper; all other taper 8°

Chart 5a

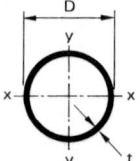

CIRCULAR HOLLOW SECTIONS

To BS4848: Part 2 unless marked †

Designation		Mass per metre	Area of section	Moment of inertia	Radius of gyration	Elastic modulus	Plastic modulus	Torsional constants	
Outside diameter D	Thickness t							J	C
mm	mm	kg	cm^2	cm^4	cm	cm^3	cm^3	cm^4	cm^3
21.3	3.2	1.43	1.82	0.77	0.650	0.72	1.06	1.54	1.44
26.9	3.2	1.87	2.38	1.70	0.846	1.27	1.81	3.41	2.53
33.7	2.6	1.99	2.54	3.09	1.10	1.84	2.52	6.19	3.67
	3.2	2.41	3.07	3.60	1.08	2.14	2.99	7.21	4.28
	4.0	2.93	3.73	4.19	1.06	2.49	3.55	8.38	4.97
42.4	2.6	2.55	3.25	6.46	1.41	3.05	4.12	12.9	6.10
	3.2	3.09	3.94	7.62	1.39	3.59	4.93	15.2	7.19
	4.0	3.79	4.83	8.99	1.36	4.24	5.92	18.0	8.48
48.3	3.2	3.56	4.53	11.6	1.60	4.80	6.52	23.2	9.59
	4.0	4.37	5.57	13.8	1.57	5.70	7.87	27.5	11.4
	5.0	5.34	6.80	16.2	1.54	6.69	9.42	32.3	13.4
60.3	3.2	4.51	5.74	23.5	2.02	7.78	10.4	46.9	15.6
	4.0	5.55	7.07	28.2	2.00	9.34	12.7	56.3	18.7
	5.0	6.82	8.69	33.5	1.96	11.1	15.3	67.0	22.2
76.1	3.2	5.75	7.33	48.8	2.58	12.8	17.0	97.6	25.6
	4.0	7.11	9.06	59.1	2.55	15.5	20.8	118	31.0
	5.0	8.77	11.2	70.9	2.52	18.6	25.3	142	37.3
88.9	3.2	6.76	8.62	79.2	3.03	17.8	23.5	158	35.6
	4.0	8.38	10.7	96.3	3.00	21.7	28.9	193	43.3
	5.0	10.3	13.2	116	2.97	26.2	35.2	233	52.4
114.3	3.6	9.83	12.5	192	3.92	33.6	44.1	384	67.2
	5.0	13.5	17.2	257	3.87	45.0	59.8	514	89.9
	6.3	16.8	21.4	313	3.82	54.7	73.6	625	109
139.7	5.0	16.6	21.2	481	4.77	68.8	90.8	961	138
	6.3	20.7	26.4	589	4.72	84.3	112	1177	169
	8.0	26.0	33.1	720	4.66	103	139	1441	206
	10.0	32.0	40.7	862	4.60	123	189	1724	247
168.3	5.0	20.1	25.7	856	5.78	102	133	1712	203
	6.3	25.2	32.1	1053	5.73	125	165	2107	250
	8.0	31.6	40.3	1297	5.67	154	206	2595	308
	10.0	39.0	49.7	1564	5.61	186	251	3128	372
193.7	5.0†	23.3	29.6	1320	6.67	136	178	2640	273
	5.4	25.1	31.9	1417	6.66	146	192	2834	293
	6.3	29.1	37.1	1630	6.63	168	221	3260	337
	8.0	36.6	46.7	2016	6.57	208	276	4031	416
	10.0	45.3	57.7	2442	6.50	252	338	4883	504
	12.5	55.9	71.2	2934	6.42	303	411	5869	605
	16.0	70.1	89.3	3554	6.31	367	507	7109	734
219.1	5.0†	26.4	33.6	1928	7.57	176	229	3856	352
	6.3	33.1	42.1	2386	7.53	218	285	4772	436
	8.0	41.6	53.1	2960	7.47	270	357	5919	540
	10.0	51.6	65.7	3598	7.40	328	438	7197	657
	12.5	63.7	81.1	4345	7.32	397	534	8689	793
	16.0	80.1	102	5297	7.20	483	661	10590	967
	20.0	98.2	125	6261	7.07	572	795	12520	1143

Chart 5b

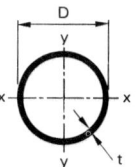

CIRCULAR HOLLOW SECTIONS

To BS4848: Part 2 unless marked †

Designation		Mass	Area	Moment	Radius			Torsional constants	
Outside diameter D	Thickness t	per metre	of section	of inertia	of gyration	Elastic modulus	Plastic modulus	J	C
mm	mm	kg	cm^2	cm^4	cm	cm^3	cm^3	cm^4	cm^3
244.5	6.3	37.0	47.1	3346	8.42	274	358	6692	547
	8.0	46.7	59.4	4160	8.37	340	448	8321	681
	10.0	57.8	73.7	5073	8.30	415	550	10150	830
	12.5	71.5	91.1	6147	8.21	503	673	12290	1006
	16.0	90.2	115	7533	8.10	616	837	15070	1232
	20.0	111	141	8957	7.97	733	1011	17910	1465
273	6.3	41.4	52.8	4696	9.43	344	448	9392	688
	8.0	52.3	66.6	5852	9.37	429	562	11700	857
	10.0	64.9	82.6	7154	9.31	524	692	14310	1048
	12.5	80.3	102	8697	9.22	637	849	17390	1274
	16.0	101	129	10710	9.10	784	1058	21410	1569
	20.0	125	159	12800	8.97	938	1283	25600	1875
	25.0	153	195	15130	8.81	1108	1543	30250	2216
323.9	8.0	62.3	79.4	9910	11.2	612	799	19820	1224
	10.0	77.4	98.6	12160	11.1	751	986	24320	1501
	12.5	96.0	122	14850	11.0	917	1213	29690	1833
	16.0	121	155	18390	10.9	1136	1518	36780	2271
	20.0	150	191	22140	10.8	1367	1850	44280	2734
	25.0	184	235	26400	10.6	1630	2239	52800	3260
355.6	8.0	68.6	87.4	13200	12.3	742	967	26400	1485
	10.0	85.2	109	16220	12.2	912	1195	32450	1825
	12.5	106	135	19850	12.1	1117	1472	39700	2233
	16.0	134	171	24660	12.0	1387	1847	49330	2774
	20.0	166	211	29790	11.9	1676	2255	59580	3351
	25.0	204	260	35680	11.7	2007	2738	71350	4013
406.4	10.0	97.8	125	24480	14.0	1205	1572	48950	2409
	12.5	121	155	30030	13.9	1478	1940	60060	2956
	16.0	154	196	37450	13.8	1843	2440	74900	3686
	20.0	191	243	45430	13.7	2236	2989	90860	4472
	25.0	235	300	54700	13.5	2692	3642	109400	5384
	32.0	295	376	66430	13.3	3269	4497	132900	8539
457	10.0	110	140	35090	15.8	1536	1998	70180	3071
	12.5	137	175	43140	15.7	1888	2470	86290	3776
	16.0	174	222	53960	15.6	2361	3113	107900	4723
	20.0	216	275	65680	15.5	2874	3822	131400	5749
	25.0	266	339	79420	15.3	3476	4671	158800	6951
	32.0	335	427	97010	15.1	4246	5791	194000	8491
	40.0	411	524	114900	14.8	5031	6977	229900	10060
508	10.0†	123	156	48520	17.6	1910	2480	97040	3821
	12.5†	153	195	59760	17.5	2353	3070	119500	4705

Chart 5c

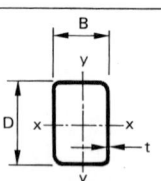

RECTANGULAR HOLLOW SECTIONS

To BS4843: Part 2 unless marked †

Designation		Mass per metre	Area of section	Moment of inertia		Radius of gyration		Elastic modulus		Plastic modulus		Torsional constants	
Size D × B	Thickness t			Axis x-x	Axis y-y	Axis x-x	Axis y-y	Axis x-x	Axis y-y	Axis x-x	Axis y-y	J	C
mm	mm	kg	cm^2	cm^4	cm^4	cm	cm	cm^3	cm^3	cm^3	cm^3	cm^4	cm^3
50 × 30	2.4†	2.81	3.58	11.6	5.14	1.80	1.20	4.66	3.42	5.80	4.02	11.3	5.57
	2.6	3.03	3.86	12.4	5.45	1.79	1.19	4.96	3.63	6.21	4.30	12.1	5.90
	2.9†	3.35	4.26	13.5	5.90	1.78	1.18	5.40	3.93	6.81	4.70	13.2	6.37
	3.2	3.66	4.66	14.5	6.31	1.77	1.16	5.82	4.21	7.39	5.08	14.2	6.81
60 × 40	2.5†	3.71	4.72	23.1	12.2	2.21	1.61	7.71	6.10	9.43	7.09	25.0	9.74
	2.9†	4.26	5.42	26.2	13.7	2.20	1.59	8.72	6.86	10.7	8.05	28.4	10.9
	3.2	4.66	5.94	28.3	14.8	2.18	1.58	9.44	7.39	11.7	8.75	30.8	11.8
	4.0	5.72	7.28	33.6	17.3	2.15	1.54	11.2	8.67	14.1	10.5	36.6	13.7
80 × 40	2.9†	5.17	6.58	53.5	17.7	2.85	1.64	13.4	8.86	16.7	10.2	42.5	14.9
	3.2	5.67	7.22	58.1	19.1	2.84	1.63	14.5	9.56	18.3	11.1	46.1	16.1
	4.0	6.97	8.88	69.6	22.6	2.80	1.59	17.4	11.3	22.2	13.4	55.1	18.9
90 × 50	2.9†	6.08	7.74	82.9	32.8	3.27	2.06	18.4	13.1	22.7	15.0	74.2	21.8
	3.6	7.46	9.50	99.8	39.1	3.24	2.03	22.2	15.6	27.6	18.1	89.3	25.9
	5.0	10.1	12.9	130	50.0	3.18	1.97	28.9	20.0	36.6	23.9	116	32.9
100 × 50	2.9†	6.53	8.32	108	36.1	3.60	2.08	21.5	14.4	26.7	16.4	85.7	24.4
	3.2	7.18	9.14	117	39.1	3.58	2.07	23.5	15.6	29.2	17.9	93.3	26.4
	4.0	8.86	11.3	142	46.7	3.55	2.03	28.4	18.7	35.7	21.7	113	31.4
	5.0	10.9	13.9	170	55.1	3.50	1.99	34.0	22.0	43.3	26.1	135	37.0
	6.3†	13.4	17.1	202	64.2	3.44	1.94	40.5	25.7	52.5	31.3	160	43.0
100 × 60	2.9†	6.99	8.90	121	54.6	3.69	2.48	24.3	18.2	29.5	20.7	118	29.8
	3.6	8.59	10.9	147	65.4	3.66	2.45	29.3	21.8	36.0	25.1	142	35.6
	5.0	11.7	14.9	192	84.7	3.60	2.39	38.5	28.2	48.1	33.3	187	45.9
	6.3	14.4	18.4	230	99.9	3.54	2.33	46.0	33.3	58.4	40.2	224	53.9
120 × 60	3.6	9.72	12.4	230	76.9	4.31	2.49	38.3	25.6	47.6	29.2	183	43.3
	5.0	13.3	16.9	304	99.9	4.24	2.43	50.7	33.3	63.9	38.8	242	56.0
	6.3	16.4	20.9	366	118	4.18	2.38	61.0	39.4	78.0	46.9	290	66.0
120 × 80	5.0	14.8	18.9	370	195	4.43	3.21	61.7	48.8	75.4	56.7	401	77.9
	6.3	18.4	23.4	447	234	4.37	3.16	74.6	58.4	92.3	69.1	486	93.0
	8.0	22.9	29.1	537	278	4.29	3.09	89.5	69.4	113	88.9	586	110
	10.0	27.9	35.5	628	320	4.20	3.00	105	80.0	134	99.4	688	126
150 × 100	5.0	18.7	23.9	747	396	5.59	4.07	99.5	79.1	121	90.8	806	127
	6.3	23.3	29.7	910	479	5.53	4.02	121	95.9	148	111	985	153
	8.0	29.1	37.1	1106	577	5.46	3.94	147	115	183	137	1202	184
	10.0	35.7	45.5	1312	678	5.37	3.86	175	136	220	164	1431	215
160 × 80	5.0	18.0	22.9	753	251	5.74	3.31	94.1	62.8	117	71.7	599	106
	6.3	22.3	28.5	917	302	5.68	3.26	115	75.6	144	87.7	729	127
	8.0	27.9	35.5	1113	361	5.60	3.19	139	90.2	177	107	882	151
	10.0	34.2	43.5	1318	419	5.50	3.10	165	105	213	127	1041	175
200 × 100	5.0	22.7	28.9	1509	509	7.23	4.20	151	102	186	115	1202	172
	6.3	28.3	36.0	1851	618	7.17	4.14	185	124	231	141	1473	208
	8.0	35.4	45.1	2269	747	7.09	4.07	227	149	286	174	1802	251
	10.0	43.6	55.5	2718	881	7.00	3.98	272	176	346	209	2154	296
	12.5	53.4	68.0	3218	1022	6.88	3.88	322	204	417	249	2541	342
	16.0	66.4	84.5	3808	1175	6.71	3.73	381	235	505	297	2988	393
250 × 150	6.3	38.2	48.6	4178	1886	9.27	6.23	334	252	405	284	4049	413
	8.0	48.0	61.1	5167	2317	9.19	6.16	413	309	505	353	5014	506
	10.0	59.3	75.5	6259	2784	9.10	6.07	501	371	618	430	6082	606
	12.5	73.0	93.0	7518	3310	8.99	5.97	601	441	751	520	7317	717
	16.0	91.5	117	9089	3943	8.83	5.82	727	526	924	635	8863	851
300 × 200	6.3	48.1	61.2	7880	4216	11.3	8.30	525	422	627	475	8468	681
	8.0	60.5	77.1	9798	5219	11.3	8.23	653	522	785	593	10549	840
	10.0	75.0	95.5	11940	6331	11.2	8.14	796	633	964	726	12890	1016
	12.5	92.6	118	14460	7619	11.1	8.04	964	762	1179	886	15654	1217
	16.0	117	149	17700	9239	10.9	7.89	1180	924	1462	1094	19227	1469
400 × 200	10.0	90.7	116	24140	8138	14.5	8.39	1207	814	1492	916	19236	1377
	12.5	112	143	29410	9820	14.3	8.29	1471	982	1831	1120	23408	1657
	16.0	142	181	36300	11950	14.2	8.14	1815	1195	2285	1388	28835	2011
450 × 250	10.0	106	136	37180	14900	16.6	10.5	1653	1192	2013	1338	33247	1986
	12.5	132	168	45470	18100	16.5	10.4	2021	1448	2478	1642	40668	2407
	16.0	167	213	56420	22250	16.3	10.2	2508	1780	3103	2047	50478	2948

Chart 6

Diagram to scale for 25mm diameter

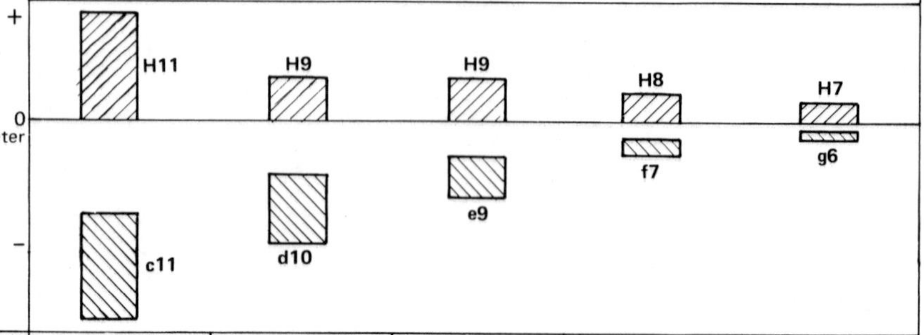

Clearance fits: H11, c11, H9, d10, H9, e9, H8, f7, H7, g6

Nominal sizes		Tolerance		Tolerance		Tolerance		Tolerance		Tolerance	
Over	To	H11	c11	H9	d10	H9	e9	H8	f7	H7	g6
mm	mm	0.001 mm	0.001 mm	0.001 mm	0.001 mm	0.001 mm	0.001 mm	0.001 mm	0.001 mm	0.001 mm	0.001 mm
—	3	+60 0	−60 −120	+25 0	−20 −60	+25 0	−14 −39	+14 0	−6 −16	+10 0	−2 −8
3	6	+75 0	−70 −145	+30 0	−30 −78	+30 0	−20 −50	+18 0	−10 −22	+12 0	−4 −12
6	10	+90 0	−80 −170	+36 0	−40 −98	+36 0	−25 −61	+22 0	−13 −28	+15 0	−5 −14
10	18	+110 0	−95 −205	+43 0	−50 −120	+43 0	−32 −75	+27 0	−16 −34	+18 0	−6 −17
18	30	+130 0	−110 −240	+52 0	−65 −149	+52 0	−40 −92	+33 0	−20 −41	+21 0	−7 −20
30	40	+160 0	−120 −280	+62 0	−80 −180	+62 0	−50 −112	+39 0	−25 −50	+25 0	−9 −25
40	50	+160 0	−130 −290								
50	65	+190 0	−140 −330	+74 0	−100 −220	+74 0	−60 −134	+46 0	−30 −60	+30 0	−10 −29
65	80	+190 0	−150 −340								
80	100	+220 0	−170 −390	+87 0	−120 −260	+87 0	−72 −159	+54 0	−36 −71	+35 0	−12 −34
100	120	+220 0	−180 −400								
120	140	+250 0	−200 −450	+100 0	−145 −305	+100 0	−84 −185	+63 0	−43 −83	+40 0	−14 −39
140	160	+250 0	−210 −460								
160	180	+250 0	−230 −480								
180	200	+290 0	−240 −530	+115 0	−170 −355	+115 0	−100 −215	+72 0	−50 −96	+46 0	−15 −44
200	225	+290 0	−260 −550								
225	250	+290 0	−280 −570								
250	280	+320 0	−300 −620	+130 0	−190 −400	+130 0	−110 −240	+81 0	−56 −108	+52 0	−17 −49
280	315	+320 0	−330 −650								
315	355	+360 0	−360 −720	+140 0	−210 −440	+140 0	−125 −265	+89 0	−62 −119	+57 0	−18 −54
355	400	+360 0	−400 −760								
400	450	+400 0	−440 −840	+155 0	−230 −480	+155 0	−135 −290	+97 0	−68 −131	+63 0	−20 −60
450	500	+400 0	−480 −880								

	Transition fits				Interference fits						
H7 / h6	H7 / k6		H7 / n6		H7 / p6		H7 / s6			Holes / Shafts	
Tolerance		**Tolerance**		**Tolerance**		**Tolerance**		**Tolerance**		**Nominal sizes**	
H7	h6	H7	k6	H7	n6	H7	p6	H7	s6	Over	To
0.001 mm	0.001 mm	0.001 mm	0.001 mm	0.001 mm	0.001 mm	0.001 mm	0.001 mm	0.001 mm	0.001 mm	mm	mm
+10 / 0	−6 / 0	+10 / 0	+6 / +0	+10 / 0	+10 / +4	+10 / 0	+12 / +6	+10 / 0	+20 / +14	—	3
+12 / 0	−8 / 0	+12 / 0	+9 / +1	+12 / 0	+16 / +8	+12 / 0	+20 / +12	+12 / 0	+27 / +19	3	6
+15 / 0	−9 / 0	+15 / 0	+10 / +1	+15 / 0	+19 / +10	+15 / 0	+24 / +15	+15 / 0	+32 / +23	6	10
+18 / 0	−11 / 0	+18 / 0	+12 / +1	+18 / 0	+23 / +12	+18 / 0	+29 / +18	+18 / 0	+39 / +28	10	18
+21 / 0	−13 / 0	+21 / 0	+15 / +2	+21 / 0	+28 / +15	+21 / 0	+35 / +22	+21 / 0	+48 / +35	18	30
+25 / 0	−16 / 0	+25 / 0	+18 / +2	+25 / 0	+33 / +17	+25 / 0	+42 / +26	+25 / 0	+59 / +43	30	40
+25 / 0	−16 / 0	+25 / 0	+18 / +2	+25 / 0	+33 / +17	+25 / 0	+42 / +26	+25 / 0	+59 / +43	40	50
+30 / 0	−19 / 0	+30 / 0	+21 / +2	+30 / 0	+39 / +20	+30 / 0	+51 / +32	+30 / 0	+72 / +53	50	65
+30 / 0	−19 / 0	+30 / 0	+21 / +2	+30 / 0	+39 / +20	+30 / 0	+51 / +32	+30 / 0	+78 / +59	65	80
+35 / 0	−22 / 0	+35 / 0	+25 / +3	+35 / 0	+45 / +23	+35 / 0	+59 / +37	+35 / 0	+93 / +71	80	100
+35 / 0	−22 / 0	+35 / 0	+25 / +3	+35 / 0	+45 / +23	+35 / 0	+59 / +37	+35 / 0	+101 / +79	100	120
+40 / 0	−25 / 0	+40 / 0	+28 / +3	+40 / 0	+52 / +27	+40 / 0	+68 / +43	+40 / 0	+117 / +92	120	140
+40 / 0	−25 / 0	+40 / 0	+28 / +3	+40 / 0	+52 / +27	+40 / 0	+68 / +43	+40 / 0	+125 / +100	140	160
+40 / 0	−25 / 0	+40 / 0	+28 / +3	+40 / 0	+52 / +27	+40 / 0	+68 / +43	+40 / 0	+133 / +108	160	180
+46 / 0	−29 / 0	+46 / 0	+33 / +4	+46 / 0	+60 / +31	+46 / 0	+79 / +50	+46 / 0	+151 / +122	180	200
+46 / 0	−29 / 0	+46 / 0	+33 / +4	+46 / 0	+60 / +31	+46 / 0	+79 / +50	+46 / 0	+159 / +130	200	225
+46 / 0	−29 / 0	+46 / 0	+33 / +4	+46 / 0	+60 / +31	+46 / 0	+79 / +50	+46 / 0	+169 / +140	225	250
+52 / 0	−32 / 0	+52 / 0	+36 / +4	+52 / 0	+66 / +34	+52 / 0	+88 / +56	+52 / 0	+190 / +158	250	280
+52 / 0	−32 / 0	+52 / 0	+36 / +4	+52 / 0	+66 / +34	+52 / 0	+88 / +56	+52 / 0	+202 / +170	280	315
+57 / 0	−36 / 0	+57 / 0	+40 / +4	+57 / 0	+73 / +37	+57 / 0	+98 / +62	+57 / 0	+226 / +190	315	355
+57 / 0	−36 / 0	+57 / 0	+40 / +4	+57 / 0	+73 / +37	+57 / 0	+98 / +62	+57 / 0	+244 / +208	355	400
+63 / 0	−40 / 0	+63 / 0	+45 / +5	+63 / 0	+80 / +40	+63 / 0	+108 / +68	+63 / 0	+272 / +232	400	450
+63 / 0	−40 / 0	+63 / 0	+45 / +5	+63 / 0	+80 / +40	+63 / 0	+108 / +68	+63 / 0	+292 / +252	450	500

Index